THE CLIMATE CHANGE DECEPTION

THE CLIMATE CHANGE DECEPTION

The Alarming Corruption of Sciene

JUSTIN JOHN EDWARD MABIE, PHD

Skyhorse Publishing

Skyhorse Publishing books may be purchased in bulk at special discounts for sales promotion, corporate gifts, fund-raising, or educational purposes. Special editions can also be created to specifications. For details, contact the Special Sales Department, Skyhorse Publishing, 307 Fifth Avenue, 4th Floor, New York, NY 10016 or info@skyhorsepublishing.com.

Skyhorse® and Skyhorse Publishing® are registered trademarks of Skyhorse Publishing, Inc.®, a Delaware corporation.

Visit our website at www.skyhorsepublishing.com.

10 9 8 7 6 5 4 3 2 1

Library of Congress Cataloging-in-Publication Data is available on file.

Cover design by David Ter-Avanesyan
Cover photo credit: Justin Mabie

Print ISBN: 978-1-5107-8642-4
Ebook ISBN: 978-1-5107-8643-1

Printed in the United States of America

To my grandfathers, John Mabie and Nicholas Squicciarini. From Peleliu to the Battle of the Bulge, they fought like hell to protect our world from tyrants. I pray that my fight on this academic battlefield will honor their sacrifice.

CONTENTS

ACKNOWLEDGMENTS

This is a story about our small community of Earth scientists.

Some of my friends and colleagues might disapprove of me telling this story. Those who know me will understand that I have good intentions, and I hope that some will be thankful for the sacrifices I've made to tell their story.

Finishing this book wouldn't have been possible without my wife, Tatiana Sazonova. She gives me the support I need when the weight of the world is too much to manage. She also gives me the kick in the pants I need when I fly off of the rails.

I apologize to my children for spending so many hours on this book. Those hours would have been better spent with them.

Thank you, Lord, for blessing me and for giving me strength and courage.

CHAPTER 1

THIS IS OUR NOAA

In the climate change debate either you "trust the science" or you're a "climate denier." This condescending oversimplification treats you like you are ignorant. You deserve to know why you are being insulted and shamed into thinking that way. This is the story of the climate change industry and the corruption of our institutions.

In 1967 the University of Colorado partnered with the US government to create a new type of pseudoacademic department called a cooperative institute. Since then, several cooperative institutes have been given broad access to our federal government, and some have used that access to improperly leverage the power and influence of our National Oceanic and Atmospheric Administration (NOAA).

NOAA was created in 1970 by Richard Nixon—who knew? In order to help protect our country from natural hazards, several existing federal functions were consolidated into the one NOAA.[1]

From 2008 until 2019, I served as the program leader for the World Data Center A (WDC-A) for geomagnetism, which evolved from one of the earliest precursors of NOAA. The program originated within the US Coast and Geodetic Survey (CGS), which was a government agency set up in 1807 by Thomas Jefferson. The CGS supported maritime safety in the time when shipping was our principal mode of commercial transportation. The CGS mission was to survey the coast, create accurate nautical maps, and measure the Earth's magnetic field—an essential task during a time when navigators relied on the compass.

Before the Second World War, scientific enterprises weren't organized in the way we think of them today. The exploits of Nicola Tesla bring to mind an extreme example of the mad scientist freelancing cutting-edge research that would inevitably lead to creation of monstrosities akin to the Frankenstein monster. The refined methods of Tesla's colleague Thomas Edison provide some level-headed context for the more formal scientific endeavors of the time. Those enterprises are very different from what we'd find at a modern university today.

The time period leading up to World War II was a great renaissance of scientific achievement that was largely driven by close collaboration between the world's greatest minds. But the war was going to be won or lost on technological advancements, and the warring governments were forced to classify and compartmentalize scientific research. The result was a divergence of standards and practices between different countries, and even different agencies within a given country. Even basic methods such as how to record data were completely different in the United States, Soviet Union, Japan, and the United Kingdom. After the war, the scientific community desperately needed to normalize how we studied and recorded observations of the atmosphere and oceans.

Scientific study of the Earth's weather and climate entered into a modern golden age in 1957 with the start of the International Geophysical Year (IGY). During the IGY, world governments agreed to normalize how common observations were made and recorded and how that data would be shared and preserved. Probably the greatest accomplishment of the IGY was that the United States and the Soviet Union agreed that understanding global weather was a shared national security interest they needed to cooperate on. Despite their geopolitical isolation during the Cold War, the two adversaries agreed to share environmental data. From that agreement the WDC-A was created to be hosted by the United States, and World Data Center B (WDC-B) was created to be hosted by the USSR. Soon thereafter, other countries would join the effort, collectively hosting World Data Center C (WDC-C).

In 1965 there was a reorganization in the Department of Commerce (DOC) that created the Environmental Science Services Administration

(ESSA), which established an overarching Earth sciences service within the US government.

The four ESSA institutes for environmental research were the Institute for Telecommunication Sciences and Aeronomy, the Institute for Earth Sciences, the Institute for Oceanography, and the Institute for Atmospheric Sciences. The science and data collection programs that I worked on and managed came from major components of these institutes. The Institute for Earth Sciences included the remnants of the CGS, including the geomagnetism program that I managed.

When my employment at NOAA was terminated, my mentor and I were all that remained of what had become of another major institute, the Central Radio Propagation Laboratory.

The story I am about to tell you is how NOAA's legacy was supplanted and replaced with something different. I will show you that the goal of that was to repurpose this once great organization and make sure that we forget where it came from and why. In its most basic sense, NOAA has abandoned its original mission and replaced it with a rogue political mission run by operatives who represent special interests.

The irony in the history of NOAA is that today's NOAA is working toward the antithesis of what Nixon's NOAA was set up to accomplish. NOAA has become a political monster that serves special interests and a manufactured climate change industry. In recent years, NOAA leadership has been working to advance the United Nations (UN) policy of carbon sequestration, which undermines our national interests. This is in defiance of NOAA's original and most important mission, which is to protect our country from the natural hazards that are expected to become more severe and more frequent as the Earth becomes warmer.

As a government institution, NOAA should not be advancing the political agenda of the UN, especially when that could undermine our national security. NOAA's real mission is to collect data, understand our environment, promote commerce, and help harden and protect our infrastructure against changing weather and climate.

In effect, the push to sequester carbon harms all major components of this mission. My story will show how opportunistic UN institutions have taken advantage of the climate change crisis to leverage power and influence over the world's governments. I will show that they have no interest in the preservation of climate data records and they don't really care about understanding the science of climate change that the data can help teach us.

By leveraging NOAA's reputation, special interests have been pressing hard to undermine our nation's commercial interests in lieu of some theoretical environmental benefit in the distant future. The same people have been promoting an unrealistic apocalyptic vision of climate change that fails to address the basic need of hardening of our infrastructure against climate change.

The global ruling class treats the American people like we are ignorant and untrustworthy. I believe the American people are inherently good and conscientious and we can restore a government that operates by and for the people. We can have a clean environment and resilient infrastructure at the same time, and our planet will be better off for it.

A Failure of Diplomacy

NOAA, an arm of the executive branch of the United States government, has been the world's most reputable climate change agency for decades. Research that has been done by NOAA, or in collaboration with NOAA, provides the foundation of our understanding of weather and climate change. Because of that, NOAA is the world's most powerful and influential organization for anything and everything that has to do with weather and climate.

The nations and empires that have wielded the most power have traditionally given us our greatest advancements in science and technology. Our modern climate record begins in the early days of the industrial revolution when understanding the weather was key to global commerce. For safe and efficient navigation, we modernized timekeeping, studied the ocean currents and the winds, created records of ocean and air temperatures, began monitoring the geomagnetic field, and created climatological maps of the Earth.

We emerged from World War II with modern technological marvels like rockets and nuclear weapons. With that, kinetic conflicts gave way to a Cold War. When President John F. Kennedy announced that we would land a man to the moon, we were under the threat of nuclear Armageddon, and diplomatic relations with the Soviet Union (USSR) were almost nonexistent.

The foundations of the modern global economy were built on being able to understand and forecast the weather. One of the mutual interests that we shared with the USSR was the need to understand and forecast the weather. As part of a multilateral effort, our nations formed the World Data Centers (WDCs) as a partnership between the United States and the Soviet Union that was designed to persist no matter how bad diplomatic relations would become. That partnership would last for a half century and lay the foundations for our modern global monitoring and research systems.

When I began working at WDC-A for Solar-Terrestrial Physics in 2005, NOAA was led by Admiral Conrad Lautenbacher. Under Admiral Lautenbacher's leadership, NOAA was a good example of how a government agency could facilitate responsible global leadership. As a natural consequence of being a government organization that is focused on weather and climate, our ranks were overwhelmingly populated with people who were politically left-leaning. That didn't seem to be a problem because they were focused on the mission. Federal and contract scientists working together were leading the world in advancements in climate monitoring, weather forecasting, modeling, and the physics of the atmosphere and oceans. NOAA's reputation as the world leader in weather and climate science was well earned and justified.

The organization was managed from the ground up, where supervisors relied on the knowledge and expertise of their data managers and scientists. This is in contrast to today's NOAA, which is managed from the top down and where politically appointed executives run the organization on behalf of special interests. This comes at the expense of the taxpayer and the mission.

When I began working at NOAA, a covert plot to repurpose and politicize the WDC was already underway. Within a decade, NOAA leadership had abandoned the mission of international cooperation and climate data stewardship. The senior people in the organization were replaced with political operatives, who focused on international efforts to influence world governments for the purpose of implementing a new system of global governance under the guise of a climate change catastrophe. The authority that has been established under this new system uses fear to consolidate power.

In 2005 Admiral Lautenbacher signed an important agreement with his Russian counterpart, Director Alexander Frolov of the Russian Federal Service for Hydrometeorology and Environmental Monitoring (ROSHYDROMET). Both men were dedicated public servants who worked hard to come to an agreement to share real-time weather data. The idea is simple: If you want to know what the weather will be like in Chicago in five days, you need to know the weather in Siberia today; if you want to forecast next week's weather in Moscow, you need to know the weather in Wisconsin today. In the same spirit of the WDC partnership, these leaders recognized the need to cooperate, irrespective of what direction the geopolitical winds were blowing. The agreement they made was brilliantly written to be renewed automatically and could only be terminated if one party made a conscious decision to withdraw from it.

In 2014 the people of Ukraine were provoked into overthrowing their government, and the United States began a proxy war with the Russian Federation.

On March 25, 2014, the director of the National Weather Service (NWS), Louis Uccellini, unilaterally terminated that agreement. He did so by cancelling a bilateral meeting with Director Frolov. He wrote:

> Dear Dr. Frolov:
>
> I regret to inform you that I am unable to conduct our teleconference on April 1 and unable to meet with you the week of May 12 for our bilateral meeting. The U.S. Government is taking this action because of the Government of Russia's recent actions in Ukraine.
>
> Sincerely,
> Louis W. Uccellini

To a typical American, Director Uccellini's letter doesn't seem to be intentionally withdrawing from an international agreement, but Uccellini is not a typical American: He was the director of our NWS, the organization that provides the weather forecasting capability that our nation relies on. Uccellini had the resources and the experience to understand that in Russian culture, he was terminating the bilateral cooperation that Admiral Lautenbacher and Director Frolov had put into place.

The director of the NWS had no authority to terminate the agreement. That decision was above his pay grade and should have been up to his boss, the undersecretary of commerce for Oceans and Atmosphere, also known as the NOAA administrator.

When pressed by Admiral Lautenbacher, Uccellini acknowledged that he knew full well what he was doing and that his actions were intentional. He made it clear that he was motivated by politics and not our national security needs to have reliable weather forecasts.

A Diplomatic Mission to Russia

I met Admiral Lautenbacher at a scientific workshop in the spring of 2014. The admiral was retired from NOAA and working in private industry. He graciously spared some time for me, and I told him about my challenges in getting real-time Russian radar data through the WDC. He told me about the agreement he made with Director Frolov and offered his help to enforce that agreement with the Russians.

After my meeting with Lautenbacher, I reached out to ROSHYDROMET and asked if we could begin a data exchange.

I was contacted by Alexander Nurullaev, director of ROSHYDROMET International Relations. Nurullaev told me he would be happy to help start our data exchange, but I would first need to visit him in Moscow.

My team decided to start our collaboration with a gesture of goodwill. Our computer systems at NOAA were set up to take in data from all over the world and push it back out to a few priority customers. Most data users would need to get data from the NOAA website or by submitting a request, but real-time customers like the NWS needed the data pushed to them the minute it was received. Because of the importance of the Russian partnership, we gave ROSHYDROMET access to our real-time system.

Once I arrived in Moscow, Director Nurullaev wrote to me to tell me that he would not meet with me:

> Dear Mr. Mabie,
>
> Thank you for your request to meet with the managers of Roshydromet. But we were informed by NOAA letter signed by Mr. Uccellini dated 25.03.2014 that NOAA has suspended the bilateral cooperation with Roshydromet under the MOU.[i]
>
> In case of new decision we would be glad to receive an official letter from NOAA informing us about readiness to work again bilaterally under the MOU.
>
> With best regards,
> Alexander Nurullaev
> Head of International Relations
> Roshydromet

I would eventually come to understand that the Russians had never intended to meet with me. They wanted me to come to Moscow, so I would take their message back to my superiors. The message was to tell the NWS director that the Russians didn't want to end their bilateral cooperation, and they needed an official letter from NOAA telling them that we still wanted to work together.

I reported back to the acting director of the NOAA National Geophysical Data Center (NGDC), Eric Kihn, and to Admiral Lautenbacher. I assume that Eric sent my message up his chain of command, but Lautenbacher wanted to take a more direct approach. The admiral had worked hard to forge that agreement, he understood how important it was, and he was a patriot who still prioritized our national interests.

The admiral's character was rare in an environment that I would come to realize was full of elitist thugs. Not only was he kind enough to spare his time for a low-level scientist, but he also treated me with respect and asked for my permission before he took any action. He asked me if he could reach out to NWS Director Uccellini about the matter.

i MOU is a Memorandum of Understanding.

The admiral's correspondence started by complimenting Uccellini on his work and then explained that he was interested in continued collaboration with the Russians. He told a story about how he had recently seen Frolov at a meeting of the World Meteorological Organization. Consistent with my experience with Russian government officials, Frolov was described as being friendly and eager to continue their partnership with NOAA.

Admiral Lautenbacher even gave the NWS director the benefit of the doubt on who could have made the decision to cancel the meeting. He suggested that whatever may have caused the "State Department" to cancel the meeting, that should not cause us to dispense with the partnership for all time. I'm sure the admiral knew or suspected that it was Uccellini himself, not the State Department, who was responsible. If he didn't already know, it would soon become clear.

The admiral identified me as the employee who was trying to work with the Russians, and he encouraged Uccellini to find a way to reengage with Director Frolov.

The NWS director's response to his former boss came across as arrogant and hotheaded. He seemed more interested in his personal convictions than his obligations to the taxpayer, and he was willing to compromise the NOAA mission in favor of a grudge he had with the Russian government's foreign policy.

Even if the State Department had directed high-ranking government officials to offend the Russians as some form of childish rebuke, the NWS director had a responsibility to protect his mission to deliver our nation the most accurate and reliable weather forecasts possible. Uccellini's obligation was to leverage NOAA's half-century-old WDC policy to collaborate with the Russians irrespective of the politics of the day.

Instead, the man responsible for your weather forecast threw a temper tantrum and abandoned a historic policy of cooperation.

It wasn't long before I received a call from the assistant administrator's office. Not only was Director Uccellini disrespectful to Admiral Lautenbacher, but he was also going to attack the messenger. Uccellini complained to National Environmental Satellite Data and Information Services (NESDIS) Assistant Administrator (AA) Stephen Volz, and asked him to have me disciplined. Volz as the AA of NESDIS was effectively the second in the chain of command at NOAA and arguably the most senior NOAA leader within the permanent government. Asking AA Volz to discipline me was a blatantly hostile act that I can't explain except as a sadistic attempt by a powerful official to destroy a low-level employee.

I explained to Director Volz's secretary that I was only doing my job and I was working with former NOAA administrator, Admiral Lautenbacher. She seemed to have accepted that explanation and didn't press me further. A few days later, Eric called me into his office because Stephen Volz wanted me to be disciplined. Since I had done nothing wrong, there wasn't much Eric could do.

What I failed to realize was that it was Eric Kihn himself who was at fault. Eric was responsible for having me, a cooperative institute (CI) employee, acting in a federal role as a WDC program leader. Conducting a diplomatic mission with the Russian government was a basic function of my WDC position, and by law, only a federal employee can conduct diplomacy on behalf of the US government.

Eric told me not to worry about it, but he was being deceptive. NOAA leadership had realized that as a contractor I was performing duties that were legally only allowed to be done by a federal employee. This made me a liability and put me in the crosshairs of senior leadership. From that moment on, my fate had been sealed. Not only would I never be given the federal position that I was promised, but also my credibility and reputation would have to be destroyed.

The Deep State at NOAA

In the government, the highest-level officials come and go with each administration. When President Barack Obama took office, he replaced Admiral Lautenbacher with his own administrator. But the NWS director and the assistant administrators are career employees who survive the change of presidential administrations and maintain control the powerful fiefdoms that are the NOAA line offices.

It is people like Louis Uccellini who run the permanent government and are leaders within the so-called deep state. The senior leadership in our federal government comprises the people who have been corrupted by a range of special interests, undermining the government's mission. Each corrupt act in of itself might not have big consequences, but by a "death of a thousand cuts," the problem becomes existential.

The story I am about to tell should help provide context for how our government institutions are controlled by special interests. Our country has been taken over by bad actors because we allow systematic vulnerabilities to persist. God willing, we will find a way to expose those vulnerabilities and take our country back.

CHAPTER 2

CLIMATE CHANGE POLITICS AND THE SCIENTIFIC PROCESS

Climate change has been widely used to create political division, and that has caused a widespread misunderstanding of climate change science. Even the best climate scientists can confuse their personal convictions with science, and the average person has little or no chance of figuring out what science can be trusted.

In response to a severe wildfire, the UN secretary-general recently stated, "The era of global warming has ended, [and] the era of global boiling has arrived."[2] Obviously, he was making a rhetorical statement, but his point was clear—he was doubling down on the idea that climate science has been settled and we have entered an era where catastrophic climate change was upon us.

What if the UN secretary-general is correct? As a tax-paying citizen, I expect my government to act to mitigate the harm that will be caused by this unfolding catastrophe. But there seems to be very little focus on mitigation and resilience. Politicians are focused on "fixing" climate change, which we've been told is possible if we reduce our dependence on fossil fuels.

Objectively, our government focusing on carbon sequestration makes no sense. The United States is only responsible for a small percentage of global carbon, and we have already done more than any other country in the world to reduce our carbon footprint. If we were given credit for how much American technology has reduced CO2 emissions in other countries, the United States would have far exceeded being a carbon-neutral nation.

Researching green technology and helping other countries reduce their carbon use are the most practical ways for us to help reduce atmospheric carbon dioxide levels. Reducing our own carbon footprint is counterproductive and is a misuse of our limited resources. That approach, which UN agencies are proposing, will not help protect the planet from an imminent climate change catastrophe.

Public figures have benefited from confusing the difference between weather and climate. Most people understand that weather is what's happening at any given time, and climate is the long-term trend that ultimately gives rise to weather patterns. But if you pay attention, the terms "weather" and "climate" are often misused. A mass-media meteorologist might have you believe that last week's extreme weather was a consequence of climate change.[3]

The problem is that even if severe weather events are stronger and more frequent because of global warming, there is no scientific method that allows us to correlate any given storm to climate change. Even an event that is so anomalous that it only occurs once every thousand years could be caused by a simple anomaly. Statistically, the coming together, or superposition, of many factors will occasionally cause a hundred-year or thousand-year storm.

In 2013 a monsoon caused severe flooding around my home in Colorado. NOAA reported that in some areas, the rainfall amounts were a once in a thousand-year occurrence. Despite it being a thousand-year flood, the weather event wasn't really that unusual.

There are thousands of unique climate zones on the Earth. Doesn't that mean that every year, irrespective of climate change, we should expect a few places to be devastated by a thousand-year storm?

The most severe weather could also occur if the planet were cooler. From a scientific standpoint, this prevents us from associating any isolated weather event with climate change. It is only the cumulative observations we make over many decades that allow us to build a picture of how climate change has influenced the weather.

That doesn't mean that we can't tell the climate is changing. Long-term observations of air and sea surface temperatures show us the climate is changing. We just need to look at a long-enough time series of data to filter out natural oscillations like El Niño and La Niña.[4] When NOAA officials imply that it's the hottest year on record because of climate change, they are misleading us.

Let's take a look at the *Global Climate Report*[5] produced by the NOAA National Centers for Environmental Information (NCEI). They acknowledge that La Niña is associated with a cool phase for global temperatures, but 2021 and 2022 were still relatively close to the all-time hottest years despite a protracted La Niña. NCEI spins the data analysis by implying that La Niña was the reason these weren't record warm years, while ignoring the fact that the presence of La Niña actually suggests the temperature should have been much cooler. That fact suggests there might be a problem with their data analysis, which should require some investigation. Either the Earth is warming so fast we no longer have relatively cool temperatures during La Niña or there's something wrong with the data or how it's been interpreted.

There has been a politically motivated effort to make us believe that understanding global temperature changes is trivial. We are expected to believe that this is a simple matter of operating weather stations and collecting the data. In reality, determining the Earth's temperature is extremely complicated, and developing a long-term climatology of global temperatures is even more complicated.

Consider what a hot summer day is like. If we put a thermometer over a blacktop parking lot, it is going to read a much higher temperature than if we put it in the shade, under a tree, near a creek. If we want to tell someone how hot it was, which temperature should we use? Do we want to brag about how crazy hot it was and use the temperature over the blacktop, or do we want to boast that it wasn't so bad and use the temperature we took in the shade? Or, do we want to do it the hard way and try to figure out the most scientifically accurate method?

Be honest; when your car tells you it's 105F, do you account for the error that's introduced because you're on a blacktop highway, or do you go home and tell your spouse that it was 105 outside? Even if we resist our human nature to boast and instead take a scientific approach, the scientific method can be applied in many different ways with each method having its own set of errors and uncertainties.

Now consider the NWS weather station at Denver International Airport, a place that was once a grassy field but is now sprawling into a concrete jungle. That concrete and blacktop create what we call a heat island, which makes the weather station measure higher temperatures than it otherwise would have. Scientists do correct for that bias, but every correction relies on assumptions that introduce error. When trying to minimize those errors, do you think data scientists are more likely to overcorrect and end up with low temperature readings, or do you think they are more likely to undercorrect and end up with warmer temperatures? Probably these errors go both ways, but there is a political incentive to undercorrect and err on the side of it appearing warmer.

A weather station only provides data at one location. If we have a weather station at Denver Airport and one in downtown Denver, how do we estimate the temperature in between? Now think about the difference in technology between today and thirty years ago—are the weather stations making measurements in the same way? Thirty years ago, we had fewer weather stations and less satellites—how do we make the long-term climate record self-consistent? Scientists use complex algorithms that contain many uncertainties to filter out trends, interpolate between data points, and make many other corrections. Scrutinizing these methods is a fundamental part of the scientific process, but in practice they are rarely challenged.

The NCEI *Global Climate Report* doesn't cite many sources, making it difficult to scrutinize. The temperature model that is used was developed, tested, and deployed internally at NOAA and was recently updated to correct a "cool bias" they claim was a flaw in the older model. It's not clear how the algorithms work, and the report doesn't address issues with the temperature records from individual weather stations.[6]

What we have is a reported global temperature model that is created by NCEI, an organization that I've seen act corruptly, certified by the World Data System (WDS) of the United Nations, that seems to have been propped up to rubber-stamp the certification of climate data records. Those global temperatures suggest that the most recent La Niña years were among the hottest years on record, despite La Niña historically being associated with the cool phase of global temperatures. We should assume those model temperature values are faulty, and we should challenge NOAA to convince us that they are not.

Former NCEI scientist John Bates blew the whistle and claimed there are political problems with the model results.[7] But Dr. Bates's allegations were turned into a political circus, they weren't properly investigated, and his reputation was attacked by the climate change thugs.

I'm willing to accept that the planet is warming, and much of the warming has been caused by humans. If correct, that would result in extreme weather being costlier and more frequent. But I'm familiar with NCEI, and I cannot trust anything that comes out of that organization. What we need is a robust analysis of how the data has been collected and corrected and how the models turn that data into the official temperature record. Scrutinizing the data and the models will help improve the temperature record and will allow us to have more confidence in what it tells us.

Even if the temperature record is correct, I do not believe that the solutions being forced on us are likely to reduce how much carbon is released into the atmosphere. Those ideas originate from corrupt global agencies that control the organizations that advise governments on climate change. Those who benefit from that don't care about climate change, its effects on the planet, and the harm climate change will cause to the planet's most disadvantaged citizens. Irrespective of the severity of climate change, it is hard to believe that the best solution is the one that coincidentally gives the most money and power to the people who run all aspects of the global climate change industry.

The people who benefit from the climate crisis are interested in power and influence. These are the people who let no good tragedy go to waste and have leveraged climate change to spawn a trillion-dollar industry that will decide where and how we live, what we eat, what cars we drive, how we think, and even what types of families we will be allowed to raise. The supposed climate crisis is a way for the global elites to gain control over the global population.

On the other hand, we do pollute and alter our ecosystem, and human activity does cause climate change. We should be able to agree that polluting our environment is not good. Pollution is the alleged cause of anthropogenic climate change, and there really isn't a rational propollution argument. Have you ever heard someone trying to justify being a litter bug?

Common sense also tells us that most people who dedicate their lives to science don't do it so they can deceive the world. For the most part, scientists are hardworking and honest professionals. Like any group of people, we are often misguided, our lives are not always on the right path, and we make mistakes. It is also self-evident that evil people pursue power and influence, and where power and influence is to be had, evil people will exploit it. In science, there is plenty of power and influence to be had, and quite obviously, it is being exploited.

The people who have exploited science now hold power and influence over the scientific process, and that gives them the power to control the science. Let's take a look at how the scientific process works, and how science is controlled and manipulated.

The Funding of Scientific Research

As a research scientist, part of my job is to find funding opportunities to support my work. The funding opportunities are always created by some interest. Contrary to what a normal person might be led to believe, there isn't an unlimited supply of money coming from governments, private companies, and billionaires, who altruistically and blindly sponsor scientific research. To the contrary, funding is extremely limited and directed toward specific objectives.

I really enjoy doing research for the military because it is geared toward a clearly defined mission. Military research is among the purest of basic research because it is largely insulated from politics and driven by a real-world need for technological advancement. The only real downside is that most military research ends up being classified, but my success is proof that it's possible to do this type of work in view of the public. On the other end of the spectrum are the politically driven funding agencies.

Perhaps the most well-known example of a politically driven funding agency is the National Science Foundation (NSF) of the US government.

Without doubt, lots of great scientific research has come out of their programs, but the NSF is a nepotistic organization that tends to give awards to the same research programs year after year. Within the NSF organizational structure, there are like-minded people who have been brought into power by political forces. Even if an investigator has an excellent proposal for research, they might be passed over in favor of a lower quality proposal that was submitted by a research group that is already an NSF insider, and has an established record of supporting the institutional agenda.

The political nature of the funding agencies makes it risky for a scientist to try to propose a research project that is not geared toward validating anthropogenic global warming, or being on the correct side of any politically charged topic. Many scientists won't even waste their time because their proposal is unlikely to be funded. This is how the funding mechanisms work, and it is one of the reasons that published science supports prevailing narratives more often than not.

The Peer Review Process

There are also issues with how manuscripts are published. The peer review process is supposed to encourage original work and promote unconstrained publication of the results, no matter how controversial those results might be. But this is not how the process works. In reality, scientists need to publish in prestigious journals to establish themselves and maintain a good reputation throughout their careers. The gatekeepers of peer reviewed journals are the editors, who are often selected through a political process. Those editors choose which peers will review manuscripts, and those peers decide if a manuscript will be published or rejected.

As professional investigators, we are expected to volunteer our time as reviewers, and I have reviewed many publications for a few different journals. Despite my area of expertise being highly specialized, the editors of the journals I have published in rarely invite me to volunteer as a referee. For example, the prestigious journal of the American Geophysical Union, *Radio Science*, has not asked me to review a paper in years, despite the fact that they have a hard time finding competent reviewers.

Good reviewers are often hard to find. Most fields of science, like atmospheric radio science, are very specialized, and there is only a small pool of scientists to choose from. This small pool of potential reviewers has to review a large volume of submissions from their colleagues, from students, and sometimes interdisciplinary researchers. Those of us who are specialized are generally quite busy, and it is not easy to volunteer the time it takes to not only read through and understand a manuscript but then also investigate the many references, cross referencing the sections of the manuscript for self-consistency and for consistency with prior literature, generate thoughtful critique for the author, and repeating the process as many times as is necessary. Clearly, there should be much more diversity in reviewers than there is, but there is an appearance that they are often picked based on political considerations.

Once scientists choose their journal, it behooves them to understand the politics of the editorial staff and by extension, the peers who will review their paper. If we submit something a reviewer doesn't like, the paper will probably be rejected. I have seen excellent papers rejected by a single, hostile reviewer with absurd critiques. On a couple of occasions, I have even written to the editor and told them a reviewer was clearly hostile, with incorrect conclusions, on an otherwise outstanding paper that would provide great value to the scientific community. On the flip side, there are many cases of inferior manuscripts being published in prestigious journals.[8] The peer review process is quite broken.

Something good scientists know is that we must understand the politics of the journal and avoid trying to publish any result that might contradict the prevailing paradigm of thought among our peers. Not only will our manuscript be rejected, but also, quite possibly, one of the reviewers will be on the next panel that will decide if our project will be funded. This is the choice between researching only what is implied we are allowed to research or maintaining our integrity and end up throwing away our career.

It is self-evident that most scientists are good, honest, hardworking people. But the process is corrupt and broken, and the cold hard truth is that scientists cannot research unpopular topics or publish results that could call into question the prevailing doctrines.

The Intergovernmental Panel on Climate Change

Even if we ignore the flaws in the funding and peer review process and assume the science is of high quality, it should be common sense that we need to separate the actual scientific investigations from how those results are packaged up and presented to the public. When people tell you to "trust the science," they are not telling you to read the latest peer reviewed publications; they are telling you to trust somebody's interpretation of the science, and that is not the same thing.

The Intergovernmental Panel on Climate Change (IPCC) is a UN body that exists for the stated purpose to advance scientific knowledge about climate change caused by human activity. But, the IPCC doesn't even purport to do that and acknowledges that it does not conduct its own original research.[9] Because scientific knowledge is advanced through original research, the IPCC does not advance the base of scientific knowledge. To the contrary, by controlling the industry and, by extension, what research is allowed to be funded, they hinder the advancement of scientific knowledge.

What the IPCC actually advances is how governments and the public understand climate change. That is an important job that wields tremendous power and influence. When an organization has that much power, we need to pay close attention to what they are doing, and we need to make sure they are not being exploited or acting corruptly. Too much is at stake to simply trust them.

In reality, the IPCC packages up select pieces of the body of scientific research to produce a summary or assessment of the state of the climate.[10] The results are used to inform governments on how to deal with climate change, providing the foundation for how the multi-trillion-dollar climate change industry functions. The IPCC that packages up science and directs how our world is governed is composed of "scientists" who are elected by UN member states.[11]

The IPCC is the internationally accepted authority on climate change, and we are told to believe that their work adheres strictly to the highest standards of peer review. We are supposed to believe there could not be improper influence, and we should follow their conclusions blindly. Of all the purported science, that which is presented by an inherently political body, for the express purpose of guiding how nations are run and how trillions of dollars are spent, is on its face the very science that we should question most vigorously.

In reality, the IPCC comprises "senior scientists," a questionable crowd of people who answer to special interests and cannot be trusted. I'm sure there are many good, politically neutral scientists who work with the IPCC, but I suspect they don't carry much influence within the panel.

Can anyone believe that there actually is a multi-trillion-dollar industry that is not improperly influenced, managed by politicians, and run by politically appointed scientists and still maintains the highest level of integrity? Maybe the industry that will control every aspect of our future lives is not corrupt, but I don't think we should wager our future on that without asking questions.

Weather and Climate Models

Weather and climate model results are often used to promote the climate change agenda. We are told that the models are predicting an imminent catastrophe. We are also told that the more aggressively we act to reduce our use of fossil fuels, the better the outcome will be. We should ask if those claims are representative of good science.

We rely heavily on weather- and climate-prediction models. Every government, industry, and individual make daily choices based on what weather and climate models tell us. And, these models have been a focal point of climate change disinformation.

At their best, models reproduce our understanding of the environment. In other words, they leverage our preexisting understanding to compute projections. In general, the models themselves should not drive our understanding of the basic science, but they are often purported to. In truth, models don't typically work well "out of the box." We build in levers that allow us to adjust the model computations to correct problems and improve their predictive ability. In other words, our models have built-in fudge factors.

There is nothing wrong with having these built-in fudge factors. Models are often necessary and effective, and the levers that improve their capabilities are a natural and essential component of them. Models give us short-term weather so we can plan our day and our week, and they give us long-term projections so we can plan for our future. Weather and climate models are among the most important tools we have to maintain a well-functioning society. With models, we are in a golden age of knowing how to live with, and adapt to, our changing weather and climate. Similar to military research, models are mission-driven, and they represent some of the best work that environmental scientists have to offer.

So, we might ask if models can predict a "tipping point" for climate change. Maybe they can. It has been asserted that models have predicted that we humans are causing it.[12] With better computers, more robust physics, and a continually improving understanding of atmospheric sciences, the models should be doing a good job. Even radical and rapid changes of the Earth's climate can probably be modeled accurately if we can forecast them. It's reasonable to expect that models will work well until something unexpected occurs.

Can models tell us how we can fix climate change? I don't think they can.

The first problem is there isn't any clear definition of what is meant by "fixing climate change." A model does not develop its own base of knowledge and cannot think for us. Even if it could, a model simply cannot deal with that which is not defined. So, if a model could accurately predict how the physical properties of the atmosphere will change with different carbon levels, we couldn't use the model results to come up with a target unless we know what the goal is. Right now, the IPCC has presented seemingly arbitrary targets that are relative to the atmospheric composition at the start of the Industrial Revolution in the mid-nineteenth century. Where did they come up with that, and what does it mean? Are they simply taking the position that we have to undo all human influence on the climate? There are eight billion people living on this planet, each having a substantial carbon footprint. What are they really asking for?

Ultimately, models are tools that we use to help forecast weather and climate. They do not teach us basic science, and they are tuned with fudge factors to make them more accurate. Because of that, they are inherently unreliable when they are used to forecast the unknown. We need to be very careful when relying on models to predict tipping points or how climate change will affect severe weather.

CHAPTER 3

THE INFLUENCERS OF SCIENCE

The climate scientists who are most vocal might not be the most academically qualified to lecture us on the subject. Even those who are qualified are often led to believe they know more than they actually do. Only a very few scientists have become the influencers for how the public understands science. Those few have become the public voice for all scientists and have used that voice to advance their political agendas in the name of science.

Perhaps the most influential scientist and climate activist is Dr. James Hansen. Dr. Hansen has worked hard to promote his views of climate change and seems to have tried to make a positive impact on the world. It is important that some of us who are experts engage in political activism. I would think this is an essential part of generating public trust in our policies and institutions. I'm not sure if Dr. Hansen's work as an activist has helped or hurt the cause he is passionate about, but I do support my colleagues who spend their time and energy trying to educate the public. What we need to be careful about is to avoid implying that our personal opinions qualify as peer reviewed science while overselling our academic qualifications.

James Hansen earned his PhD by studying the Venusian atmosphere.[13] I find planetary atmospheres to be an interesting field of science that can provide a great amount of insight into climate change on Earth. I have even proposed leveraging the physics of the Venusian atmosphere to help understand the seismology of Venus. However, a scientist whose formal education had little to do with the Earth's climate is not off to a good start as a leading expert in the field of climate change.

To my understanding, Dr. Hansen is considered our most prominent climate change scientist. The college degrees that he has been awarded are in physics, mathematics, and astronomy, and he has no formal education in climate sciences. It was Dr. Hansen's activism, not his academic background, that has made him famous. He was a NASA space scientist who was able to pose as a climate science expert simply because he was an institute director who oversaw climate change research programs as an administrator. This is a common sleight of hand used by political operatives to misrepresent their qualifications.

Hansen used his position at NASA to gain notoriety as an expert climate change witness during a congressional hearing in 1988. In that hearing he claimed we had 99 percent confidence the Earth was in an actual warming trend and that it wasn't merely the result of natural variations. That was an outright fabrication—not dissimilar to the UN secretary-generals more recent claim that "the era of climate boiling has begun."

If that doesn't convince you, Hansen's history of political activism, including being arrested along Daryl Hannah and other celebrities in 2009 during a coal mining protest, should lead us to question Dr. Hansen's motives.

To his credit, he effectively leveraged his fame to advance his career as a scientist. But when a scientist is engaged in activism, he is not being a scientist. Likewise, when an activist is doing science, he is obliged to leave his personal baggage at the laboratory door.

The fact that Dr. Hansen is arguably the world's most influential climate scientist is a credit to his hard work. At the same time, his fame demonstrates clearly how science can be hijacked. He has hundreds of publications in fields of science that he hasn't formally studied. I have difficulty believing that anyone can generate quality peer reviewed publications across many disciplines of science, every few months, decade after decade. It's even harder to believe that same person did so with no formal education in his field of expertise, while directing prestigious laboratories and moonlighting as a political activist.

What I am trying to convey is that his work as a scientist should not advance his legitimacy as a political activist. Likewise, his work as a political activist should not give his science more credibility. In his role as an activist, he is not a scientist.

When NASA blocked a media request for an interview, Hansen filed a whistleblower complaint claiming they tried to muzzle him. This is very different than the story I'm about to tell where I had to file a whistleblower complaint after being threatened for protecting climate data records. I had always understood that NOAA had the right to stop me from speaking on behalf of, or publishing on behalf of, NOAA. This was never an issue because I always maintained the right to speak and publish through my university affiliation.

Even though NASA admitted to being at fault, Dr. Hansen's complaint was probably misplaced to begin with. My story will show that federal investigators give credence to phony complaints when it's politically expedient—such as when a famous "climate scientist" is allegedly muzzled by a Republican administration. But when real and severe fraud, waste, and abuse occur, those same federal investigators will become complicit in the cover up to protect the deep state.

Dr. Hansen's media request could have easily taken place outside of his government role. Certainly, he had a right to present his results to the public, but he had no right to do so on behalf of the American people. Besides that, a NASA employee failing to satisfy a media request is not the same as the NWS director attacking a low-level employee for trying to get data from the Russians. It is fair to ask if Dr. Hansen's whistleblower complaint was legitimate, or was it politically motivated? Winning the fight did not in of itself make his claim legitimate.

In this book, I am telling my story. I am not publishing scientific results, and no one should ever reference it as if it were a scientific publication. What you are reading is a critique of the scientific and climate change industries, not a work of science itself. We need to be more careful in deciding what we understand to be science and what we understand to be an agenda presented by scientists.

A Complex System

Humans live in the troposphere, which is the lowest part of the Earth's atmosphere. For the most part, the public's understanding of climate change is limited to what happens here. We think in terms of how hot it is, how much it rains, and the frequency and strength of severe storms. Even most scientists tend to think in those terms, mostly ignoring other important components like the oceans, the upper atmosphere, the solid Earth and the Sun. Of course, atmospheric scientists understand how these things contribute to our weather and climate, but as human beings it is natural to simplify things. The reality is that the Earth's climate system is so complex that it's not possible for any of us to have a comprehensive understanding of all of its components.

Most climate scientists are only vaguely aware of the fact that our atmosphere actually extends quite far into "space," which begins about fifty miles above the ground.[14] Most people would be surprised to know that the Earth's atmosphere actually extends for tens of thousands of miles in altitude. Isn't it an important fact that all of our weather originates from the Sun and that the Sun's energy has to pass through those thousands of miles of the Earth's upper atmosphere before affecting our weather and climate? As an expert in planetary atmospheres, Dr. Hansen is one of the few well-known climate-change activists whose credentials suggest he should understand this well.

On the other hand, the oceans and the solid Earth are also key components of our weather and climate. Does Dr. Hansen's formal academic background cover solid Earth geophysics and oceanography? What qualifies him to be an expert on sea level rise due to ice melt? I'm sure he has a good interdisciplinary understanding of the Earth's climate, but if he is the world's leading climate change expert, we should ask how many hours he has spent in a classroom studying the many disciplines that must be mastered in order to truly understand climate change. The reality is that professional scientists typically specialize in a single field of science and rarely have a robust academic background that encompasses the many diverse fields of atmospheric sciences. In other words, even the best scientists don't know as much as we are led to believe.

Even beyond that, the effects of climate change are different everywhere we look. "Global warming" implies the average temperature is increasing near the surface of the Earth. When people talk about the imminent climate catastrophe, they might point out severe weather events like typhoons or wildfires, but they typically won't highlight what you will experience in your very specific geographic location. In general, they don't know. We can look at the historical data and the models, and maybe we can figure that out. The scientific "expert" you hear yammer on the TV, probably has no clue what you will experience in your hometown.

In simplest and most general terms, I view the energy in the Earth's atmosphere like a pendulum. If you put more energy into the pendulum, the bob will swing further. Likewise, if there is more energy in the Earth's atmosphere, we should expect the extremes to be more pronounced. Common sense would lead us to believe that the hottest heat waves will be hotter, the coldest cold snaps will be colder, the strongest storms will be stronger, and on and on. Over time, this becomes statistically significant, and we should be able to establish a cause-effect relationship.

This does mean that delicate ecosystems, like coral reefs, will be threatened by temperature extremes and are at risk. This means that ocean currents driven by winds and salinity levels will be affected by storm patterns and ice melt, and it means the frost depth will change and your pipes might freeze in the winter. Most likely there will be a large economic and human cost from climate change, and we need to be ready to deal with that. If the rate of global warming is accelerating, we should expect that cost to increase.

The prevailing theory is that greenhouse gasses emitted by burning fossil fuels are preventing the atmosphere from radiating excess heat into space; this is known as the "greenhouse effect." We can apply what we know about science, model the effects of climate change, and have some confidence that human activity is causing the Earth to warm.

The problem becomes more complicated when we consider other factors like natural cycles. It could be that the Earth is in a natural cooling cycle and manmade warming is offsetting that. Another possibility is that the natural cycles are also in a warming phase and the combined effect is a rapidly warming Earth. Maybe when the Earth warms, it finds other ways to deal with the extra heat. In all cases, basic physics tells us that polluting the environment in the way we have been doing should have a net warming effect. That is the common-sense conclusion.

Anyone who has watched a pot boil knows that water holds a lot more heat than air. There is a huge amount of water in the ocean, and if the Earth's atmosphere is warming, the ocean surface is warming with it. Basic ocean dynamics tells us that ocean currents are going to trap that heat for a long time before eventually pumping it back into the atmosphere. That means any human-caused warming from the past could eventually become a source of warming in the future. We could possibly be in a period when anthropogenic (human-caused) warming is coupling with a natural warming cycle, while the ocean is also pumping out excess heat from fifty years ago.

We might ask what happens if the Sun decides to get in on the action and becomes more active. Could the climate become much warmer overnight and become similar to when the dinosaurs roamed the Earth? That is possible and could be far more severe than the direst predictions of the IPCC.

So, this leads us to a major problem with the proposed solutions for controlling carbon emissions. Reducing our carbon footprint could help slow global warming, but even the IPCC admits that with the strictest measures they propose, it will probably not solve the problem. With current technology, windmills and solar farms, which take a tremendous amount of energy to produce, deploy, and maintain, are probably not going to make a difference.

The only realistic ways to substantially mitigate our carbon footprint seem to be to drastically reduce the world's population and to transition to a truly green and limitless energy source like hydrogen fusion power. Historically, fascists have a reputation for being aggressive and having a stomach for mass murder. It would be wise to take the threat of mass population reduction or eugenics seriously. Developing a truly green and limitless source of energy would require an effort on the scale of the Apollo project and would have no guarantee of being successful. Right now, the scientific research that is needed to develop fusion power is being slow-walked by the very same global elites who are profiting from the climate change industry.

The basic fact is that the international community is not ponying up the trillions of dollars that is needed for the investment in a fusion-based, zero-carbon, unlimited energy source, and they are not giving us any better ideas. If the world was serious about carbon mitigation, every cent that is spent to subsidize current forms of alternative energy could be redirected toward that goal. Even if the effort failed, we would come out of it with tremendous advances in technology, and we wouldn't be left worse off.

At minimum, if the world was serious about carbon sequestration, the government advisory committees would be proposing practical solutions. A practical approach would aim to do the best we can without upsetting the global economy. The current approach of reducing carbon at all costs is conclusively counterproductive. Ultimately, that approach will cause more harm than good.

The Climate Change Fascists Strike Back

The big story is the covert formation of a global climate change empire. In recent years, those who wish to control us have perverted the English language to obtain their ends through rhetorical force. If you google "fascist," you might even find a picture of a bad orange man. To help get ahead of the likelihood that my language will be manipulated, I turn to my handy 2001 edition of *Webster's Encyclopedic Unabridged Dictionary*:

> Fascism:
> A governmental system led by a dictator having complete power, forcibly suppressing opposition and criticism, regimenting all industry, commerce, etc., and emphasizing an aggressive nationalism and often racism.
>
> Fascist:
> 1. A person who believes in or sympathizes with fascism. 2. A member of a fascist movement or party. 3. A person who is dictatorial or has extreme right-wing views.

Notwithstanding the apparent absence of a dictator, the people who are leveraging climate change to obtain power are quite clearly fascist. The climate change fascists seek complete power over every aspect of our lives, they forcibly suppress opposition and criticism, they regiment industry through various forms of social credit and subsidy programs, they control global commerce, and they emphasize an aggressive form of what can only be considered global climate nationalism through the one-world government of the UN. Even racism is readily apparent in their policies with widespread condemnation of so-called white privilege, exacerbation of poverty in places like Africa through skyrocketing fuel prices, displacement of migrants from Latin America and North Africa, and special economic advantages given to the "developing" nations of China and India. The dictatorial form of one-world government that is being pushed on us is antithetical to modern liberal ideologies and can only be described as ultra-right-wing. The fact that many of their foot soldiers identify as progressive is irrelevant.

We should also keep in mind that economic warfare is still warfare. The effort to control and manipulate the global economy is no less harmful or deadly than a kinetic war. Much of their behavior seems to come right out of the Nazi playbook, and their enforcers employ methods reminiscent of the Gestapo.

Political scientists might condemn my use of the word "fascism," but I challenge them to come up with a better label. I have made my meaning clear, and thus I cannot be accused of misusing the language to thwart our discussion. Herein, these power brokers who will instill and leverage a fear of climate change to control us will be referred to as the climate change fascists.

One invariant is that the climate change fascists viciously attack anyone who tries to expose truth that undermines any of their ideals or institutions. These people are mean and dishonest, and they are not interested in correcting any of the very real problems we face. Take note that they seem to insist they are always right, and they label their opponents as corporate shills employed to protect the evil oil and gas industry. In other words, their arguments are based on moral superiority and not truth.

A respectable scientist, Dr. Patrick Brown recently admitted that he had omitted key points in a paper on the effects of climate change on wildfires.[15] In particular, he was careful to include only results that showed how climate change increases wildfire risk, while omitting factors that we as a society can control. Factors we can control include building and zoning standards, management of combustibles in the forest, and better regulation of power lines and other infrastructure that can start fires. The author was concerned that anything that could lead us to conclude that spending money on anything other than carbon emissions wouldn't fly with the journal editors, and he omitted those findings. In response, Dr. Brown's reputation was viciously attacked.

As somebody who lives in an at-risk area for forest fires, this caught my attention. At this moment, my home is in a canyon forest that is full of beetle-killed lodgepole pine trees. We have also just come off an unusually wet summer, and the seasonal combustibles like grass are far more abundant than normal. A few weeks of dry weather could put my home and my family at risk of a firestorm that could sweep through here in minutes. In this canyon, there is only one viable evacuation route through a tight two-lane-wide canyon road. I have spent hundreds of hours removing forest-fire fuel from my estate, keeping the grass cut, and taking all precautions I could think of to protect my home. My canyon will not burn one day because of climate change, but because it is in a forest. No matter how much the climate changes, forests burn from time to time. When my neighbors are told that it is only climate change that they need to focus on, not forest management, a disservice is being done to them that puts us all at risk.

It's not hard to see many of the ways our systems and institutions are flawed. You would have to assume that human beings are not exploiting any of these inherent flaws, and ignore troves of evidence to the contrary, to believe that the scientific process is not corrupted. It should be easy to conclude that dishonest people in positions of power and influence are undermining those of us who strive to produce the best and most complete scientific work.

Attacking the Scientist

During my graduate studies, I had the honor and privilege to work under Dr. Syun-Ichi Akasofu, who is considered by many to be the father of modern space physics. When Professor Akasofu was a young graduate student in Fairbanks, Alaska, he lived in a cabin with no running water. On cold Alaskan winter nights, he would go to the outhouse and watch the aurora borealis. He developed a novel theory on the dynamics of the aurora that wouldn't be validated by satellite observations until years later. That theory formed the basis for our understanding of how the Sun interacts with the Earth's atmosphere.

If you were excited when the aurora was visible across the country, you should thank Dr. Akasofu for discovering the physics we relied on to predict it. We should also thank Dr. Akasofu for bringing together the governments of Japan and the United States to create the International Arctic Research Center (IARC), arguably the world's leading research institution for studying climate change in the arctic.

Professor Akasofu spent his career promoting innovative science and encouraging young scientists to think outside of the box. He regularly gave lectures referencing Thomas Kuhn's book, *The Structure of Scientific Revolutions*,[16] teaching his students that they should not accept any scientific paradigm as an absolute. Dr. Akasofu taught us how to be responsible, free-thinking scientists. That was his sin.

While Professor Akasofu was director of the IARC, he made the mistake of giving a lecture at the NOAA David Skaggs Research Center in Boulder, Colorado, which was attended by a handful of self-righteous IPCC scientists who worked at the facility. The way they treated him was a travesty that I witnessed firsthand and forever changed how I look at the politically appointed scientists who work for the IPCC.

After giving an unimpeachable lecture on solar magnetic flux rope theory, Professor Akasofu presented some interesting results of climate studies made at the IARC. He went out of his way to make it clear that he didn't view himself as a climate change scientist, and he had no intention to get involved in the climate change debate. The professor was only suggesting that it was important to look at the whole body of scientific evidence, not just that which fits the prevailing global-warming paradigm.

Dr. Akasofu had been made a pariah in the climate change community for questioning the climate change "consensus" and taking research funding from oil companies. It is true that the IARC does research that is funded by the oil and gas industries. IARC scientists solicit funding from a wide range of sources, and the North Slope of Alaska is home to some of our nation's most important oil and gas reserves. The IARC should be working with the oil and gas industry. All scientific research is funded by special interests, and that is not a reason to delegitimize the professor's findings.

The professor was making a point that this is a big world and what is happening in one place is not likely what is happening everywhere else. In a warming climate, there are some observations we have made that don't quite make sense. Studying things that don't make sense is how we learn about the complexities of our physical systems. The reality is that the data complicates the "science" and is inconvenient. After all, the science is "settled," so why would we want pesky data complicating the discussion? That is why climate change fascists hate data.

But this is the point the professor was trying to make: We need to question science. And for that he was mercilessly berated by people who stood up, claiming to have the moral authority and know better because they were authors of the most recent IPCC report. Professor Akasofu tried to reiterate that he was only suggesting they look at the data and was drawing no conclusions whatsoever, but they continued to throw the metaphorical tomatoes at him.

This was a most shameful attack against a man who was, in my opinion, entirely correct. Most shamefully, these "scientists" were attacking a colleague because they couldn't handle looking at data. These climate fascists are politicians, not scientists when they are behaving in this way. These people are not interested in truth or discourse. They demand only that you shut up and take their word that the science has been settled and sequestering carbon is the only way to save the world from imminent destruction. Control of influential government organizations like NOAA is essential to the mission of the climate change fascist.

CHAPTER 4

THE PUBLIC SERVANT

Following my dustup with the director of the National Weather Service and NESDIS assistant administrator, there was a catastrophic breakdown in workplace decorum. It would be remiss of me to avoid admitting my part of the responsibility for that.

The truth is that I refused to back down in the face of bullying, I refused to allow people to abuse their positions of power, and I refused to compromise on my core principles. My refusal to back down enflamed the situation, but in doing so, some of the bad actors would expose themselves.

Within the federal bureaucracy, this sort of behavior is not tolerated. I fought back as a matter of honor and principle, knowing what the likely outcome would be. Before I tell you my story, I should tell something about how I became who I am, so you can understand why I chose to fight.

My mother's father was Nicholas Squicciarini, a first-generation Italian American, a businessman, and a war hero. He married my Nanna on December 6, 1941, the day before the Japanese Empire attacked Pearl Harbor. He joked that two wars began that weekend.

At the Battle of the Bulge, Sgt. Squicciarini was in the officer's tent drinking his morning coffee when a private ran over yelling, "The Germans are coming." But all was quiet, and he decided to walk to the front lines to see what was going on. Halfway there, walking through an open field and without warning, he was hit and found himself bleeding on the ground.

At that moment, he said that all hell had broken loose. He looked up and saw Nazi soldiers pouring through a fence line, so he picked up his rifle, and started firing. One shot each, the enemy soldiers fell, round after round, clip after clip.

Later that day, his uniform shredded from his body, in the cold winter air, he wrote a letter to his beloved wife. He wrote in part, "I don't know how many Jerrys I killed, but it sure as hell was a lot."

The lesson he taught me was that when you're knocked down, facing insurmountable odds, you get up and fight.

My other grandfather, John Mabie, was a lifelong civil engineer and Freemason. When I was a young boy, he taught me how to survey land, which turned out to be a valuable skill to have throughout my life.

Grandpa Mabie was also a war hero. I just didn't know that because he never talked about the war. He was a soldier in the First US Marine Division, which is known to have engaged in some of the most brutal combat of our country's history. He instilled in me humility and a drive for public service.

I was raised to be a proud, patriotic American. Both of my grandfathers taught me the importance of hard work, integrity, and service. They taught me to fight the smart fight and never back down. They helped me understand the importance of our family history and of passing our blessings on to future generations.

Physics

My public high school was full of remarkable educators who inspired their students. I had a passion for chemistry, which I decided to pursue at the University of Miami. There, I joined a fraternity and quickly failed out of school.

I moved back to New Jersey and started attending Ramapo College. The chair of the physics department was Theodore "Teddy" Halpern, one of the college's most highly respected professors. Teddy was on a mission to recruit physics majors, and I was an easy target. Teddy built a family of professors and students who still keep in touch after a quarter century. With their mentorship and friendship, my academics improved every year until I graduated, and I was inspired to continue my education in a new adventure.

Naively, I shipped off to graduate school believing that all educators were passionate about their profession. I was soon to find out that some are actually quite terrible teachers or couldn't care less about students and academics. At the University of Alaska Fairbanks (UAF), I learned the next set of important life lessons.

Into the Wild

I began a PhD program in space physics in August 1999. My classmates had a mix of interests across the physics disciplines of the upper atmosphere. What we were surprised to find out was the department faculty were coming off of a recent civil war over use of and access to the infamous High Frequency Active Auroral Research Project (HAARP). The more ruthless side won the fight and ran the losers out of the department.

Some of the professors went to other departments, but most uprooted their families and went off to find greener pastures. The students also suffered the consequences of the dispute. Almost all of my classmates were forced out of the university without the degrees we came for.

Despite department politics, the education I received at UAF was excellent. I met my wife, found my faith, and became married to both. Because of my thirst for outdoor adventure, I fell in love with atmospheric sciences.

In the high, barren mountains above Summit Lake, Alaska, I found God. My parents were believers, but they never pressed religion on me. Instead, they told me I needed to find it for myself. In my undergraduate years I identified as being agnostic with an open mind. Because my undergraduate physics department was very much like a family environment, we were able to discuss sensitive matters like faith. I became convinced there were three types of physics majors. I was among those who studied purely to obtain an understanding of our physical world. There appeared to be another group who identified as atheists and were determined to study physics as a means to reinforce their beliefs. The third group were the faithful, who were determined to study physics to reinforce their beliefs. All seemed to find success in their endeavors.

Although agnostic, I actively searched for answers. I prayed occasionally, and I investigated correlations between the physical and the metaphysical. I was inspired by a class I took on the history of scientific ideas, where we were taught the early Greek philosophers had a view that the universe was created out of a separation of the heavens and the earth, or in other words, a separation of the physical and the metaphysical. That explanation seemed quite analogous to a phenomenon in quantum physics where a particle and an antiparticle could be formed out of nothing, so long as the sum of their parts was equal to zero. That seemed to make more sense than the so-called Big Bang where all of the laws of physics are violated at some theoretical starting point where everything was at-once, created from nothing.

The beauty in the theory was that, as a first principle, those things of Earth are separate from the divine. As a matter of my faith, I believe wholly that it is not possible to use science to prove or disprove anything that is divine. What is inherent to one is inherently not to the other. From that, I could conclude that there is no possible means to investigate religion through science, and it is probably foolish to try to explain science through religion. In my mind, that dispelled any and all controversy on the matter.

With this logic, there is no contradiction between any laws of physics and the existence of the creator. I had a completely open mind, but I also had no expectation that any divine truth would ever be revealed to me.

One day, above Summit Lake Alaska, high in the mountains, I ventured out across a glacier on my snowmobile. This was an infinitely stupid thing to do, but I tended to do infinitely stupid things. I am certain I noticed a crevasse or two that I glided over, and I had a vague sense of the danger of the place. I had been reading John Krakauer's *Into the Wild,*[17] about a young Chris McCandless who ventured into the Alaska wilderness and perished because of a simple mistake. I knew I was in a dangerous place, but I didn't fully understand what those dangers were. Simply put, I was ignorant and foolish.

That same weekend, on a nearby glacier, one of my classmates fell into a crevasse and died.

On another day, near the same place, riding through a snowy field well above tree line, an unexpected and unlikely happening would start my path toward becoming a climate scientist. On that occasion, the snowpack was quite low for the time of year, and my snowmobile ski became lodged under a rock, and I was thrown from the sled. My snowmobile was totaled, but I was not hurt.

In the following years, I took great care to make sure there was an adequate snow pack before venturing out, and I began studying the weather circulations and precipitation patterns in the Alaska Range. I wanted to understand how global climate oscillations affect my favorite pastime, and my love for the study of climate change was born.

On another day, near the same place above Summit Lake, the divine truth was revealed to me. I was plowing through a white field of snow, with no one else around. I was truly in one of the most remote places imaginable, doing what I loved most, when the Lord revealed his presence to me. I cannot explain how it was revealed, only that it was, and that it was like a light switch had been suddenly turned on. In an instant, I went from being an agnostic to a man of faith.

On another weekend, we holed up at the Paxon River Lodge just south of Summit Lake. The lodge only had a dozen or so rooms. In the Alaska mountains in January, the only guests were sled heads like myself. The lodge owners had a large dining room, and each night they would prepare a dinner of their choice for the guests. That night, we were accompanied by a group of four fellow snowmobilers. The next day, we enjoyed one of the most magnificent powder days I could remember. The crisp, fifty below zero air blowing through our masks, accompanied by a blast of frozen powder in our faces when our machines would plow through piles of snow. All of God's majesty was ours to revel in, and we knew how unique and special our blessings were.

That night, we found that the other group ventured out, only to be caught in an avalanche. Three of the four of them were killed.

The Alaska Range above Summit Lake is a remarkable place of desolate beauty. From the top of the mountain ridges, nothing can be seen but pure white mountains. There are no trees as far as the eye can see, and when the snow pack is at its deepest, there is nothing but a pure white wilderness, and maybe the sign of a rock ridge near the top of the highest peaks. It would be like I imagine the surface of the Moon or Mars would be like if it froze and became covered in snow. And, it was probably not any safer than recreating on Mars or the Moon in a snowstorm. In this most remote place, my soul would be reborn with every visit.

After my fellow students and I were pushed out of our graduate programs, I was hired to work as a research assistant at the International Arctic Research Center that was right next door to the Geophysical Institute. I continued to research how the weather in Alaska is influenced by climate change, focusing on climate variations that occur over roughly ten-year timescales.

The most profound way climate change affected my life was through these decadal cycles. The phase of the Pacific Decadal Oscillation was the key factor in predicting how cold it would be and how much snow we would get.

It wasn't long before my wife was awarded her PhD and we were expecting our first child. We moved to New Jersey, and I left science to work as a traffic-signal electrician at the Department of Transportation. I loved being in public service, but the job was too dangerous, so after two years I decided to return to the laboratory. We moved to Colorado, and we both took jobs at NOAA.

At NOAA I supported the study of the upper atmosphere. Relatively uninformed, I believed that senior scientists were all-knowing. In reality, scientists tend to specialize in a narrow field of research, and they're not usually experts in much beyond what they learned in college. Probably most of us can't even recall what we learned in the classroom.

I realized the natural sciences are hamstrung by a lack of interdisciplinary researchers. Few of my colleagues work across several disciplines, and none had a comprehensive background in Earth systems science. I returned to school to study atmosphere and ocean sciences, so my formal education would encompass the ocean depths to the surface of the Sun.

The Public Servant

I am a weather and climate scientist who has nearly two decades of public service under his belt. When performing field work, I build and maintain Doppler radar systems called ionosondes. Similar to the Doppler radars that help forecast your local weather every day, the systems that I build and maintain provide real-time weather data. The main difference is my radar systems observe the upper atmosphere and are mostly used for radio communication and satellite-related technologies.

In the laboratory, I manipulate the radar systems to make unique experiments. This scientific work eventually led to completion of my dissertation.[18]

My work also includes signal intelligence and weather modification. I am one of the few scientists in my field who successfully made a career without obtaining formal government security clearance. On paper, that probably made me one of the most dangerous scientists in the world. On the other hand, the ability to speak freely had its advantages.

I rarely feel like I'm the smartest person in the room, and I never want to be. I think I'm a good scientist, but I'm probably not going to be on the short list for a Nobel Prize. I view myself as a typical scientific investigator.

I began my professional career as a first responder with the New Jersey Department of Transportation (NJDOT), repairing the electrical infrastructure of state and interstate highways in and around Newark, New Jersey. At the peak of my responsibilities, I was the lead electrician for more than fifty highway intersections, in one of the most dangerous places in the world for highway workers. We were under constant threat from traffic accidents, severe weather, and crime. Several times a week, I would answer an emergency call in the middle of the night, usually heading out alone, without proper safety backup, to repair traffic signals that were damaged in accidents. It was a dangerous job, but it was important work that I was proud of.

My understanding of physics was a constant reminder of any number of terrible things that could happen to me at any moment. I was looking for an opportunity to get off the highway and advance my career. When I started working at NJDOT I was assigned to the construction crew, where we installed poles and pulled electrical cable. Within months I was promoted to ladder truck, which was a technical job reserved for the brightest, most skilled, and most highly motivated. We were also approaching a period when multiple senior supervisors were getting ready to retire in our region, and I was on a fast track to be promoted to foreman and possibly as high up as regional supervisor. By now, I could have been an assistant commissioner.

I felt a need to get back into science, and it was very hard to justify the dangers of highway work when I had a family and many other opportunities that I could pursue. I began searching for federal government jobs and found one that fit my education well. The position was for a space physics data manager at the NOAA National Geophysical Data Center (NGDC). I had known about and relied on NGDC services throughout my graduate studies in Alaska, and to me, the place was legendary. I thought this could be the career opportunity of a lifetime and a chance to provide a better life for my family in beautiful Colorado.

When the call came in for the interview, I could not have been more excited. I was interviewed by Eric Kihn, who was known for development of the first online interactive database for space weather,[19] and by Dave Clark, the deputy director of NGDC, who was also well known for his work with the United Nations (UN) through the International Council of Scientific Unions (ICSU).[20] After the interview, Eric told me they wanted to hire me but the federal job was already targeted for Rob Redmon, a Cooperative Institute (CI) employee they wanted to transition to the federal service. Eric asked me if I would be willing to take Rob's old position at the CI.

This put me in a tough place. It was an opportunity to get back into science, give my family a better life, and get away from the extremely dangerous job in Newark. On the other hand, I would lose my permanent position as a civil servant where I was on track to rapidly ascend the ranks. But the opportunity was too good to pass up.

I told Eric that I was a civil servant at heart and I would take the job if he promised to bring me into the federal service as soon as possible. He agreed and promised that if I was to stick around, he would open a federal position for me as soon as he could. Relying on that promise, I accepted his offer, uprooted my family, and moved to Colorado to serve the American taxpayer as an NOAA scientist.

When I first arrived, Eric put me in charge of a suite of three climate models of the upper atmosphere, collectively called Space Weather Analysis (SWA). These models were the Global Ionosphere Thermosphere Model (GITM), which coupled the high-altitude neutral atmosphere with the plasma of the ionosphere; Assimilative Mapping of Ionospheric

Electrodynamics (AMIE), which was a model of the electrical properties of the upper atmosphere; and the Simple Inner Magnetosphere Model (SIMM), which was a model of the particle energies and densities trapped in the Earth's magnetic field in space. My job was to collect and process data to assimilate into the models, run the models on NOAA computers and supercomputers, and process and archive the output.

I was hired near the start of an extended federal hiring freeze. Within several years, almost all of the federal data managers at NGDC had retired, and management was telling us that they were not allowed to reopen their federal positions. What I would eventually learn was they were plotting to get rid of the World Data Centers and they were avoiding hiring any new federal employees for WDC-A positions.

Eric approached me and told me that Sue McLean, the division chief for the NGDC Marine Geology and Geophysics division (MGG), had a personal interest in making sure the WDC-A program for geomagnetism was managed by someone competent. I was offered and accepted the position as program leader for WDC-A for Geomagnetism. Eric told me this was a federal position that would fast-track my transition into federal service.

As a WDC-A program leader, my primary duty was to manage the ingest and archive of data and make the data available to the public. This job was a natural extension of my previous work in data assimilative modeling. I also took on the role of WDC liaison, establishing and maintaining relationships with our foreign partners and governments, so we could maintain and improve the international flow of environmental data. The international partners I worked most closely with were the Japanese, the Russians, and the British.

Part of the job was taking responsibility for the NOAA historic library of geomagnetism. Geomagnetism is the study of the Earth's magnetic field, historically concerned with the direction a compass will point. Our library included rare and unique publications, containing data dating back to the 1850s, and included the world's most extensive climate record of geomagnetic variations. When the modern study of geomagnetism was born, global commerce revolved around maritime traffic. The study of the Earth's magnetic field was of great importance for navigation, to keep ships and their crews safe and on route. The NOAA library was a remarkable archive that attracted some of the world's most renowned climate scientists.

Very soon, I was well known within the community of the International Association for Geomagnetism and Aeronomy (IAGA), as the program leader for WDC-A for Geomagnetism.

One of the first WDC tasks I was asked to tackle was to modernize our data services and products. The WDC legacy dated back to the 1950s, and many of the established methods did not translate well to modern technology. Two of our oldest and most widely used products were monthly data bulletins that had historically been printed and physically mailed to subscribers. A few years earlier the bulletins had been transitioned to digital publications that we delivered over the internet. The *Solar Indices Bulletin* (SIB) provided important data on the state of the Sun's activity. Likewise, the *Geomagnetic Indices Bulletin* (GIB) provided a summary of geomagnetic activity. Both publications were widely used by scientific researchers, government regional warning centers like the NOAA Space Weather Prediction Center (SWPC), and private industry.

Every month, it took two of our federal data managers dozens of hours to collect, organize, and analyze the necessary scientific data and generate the bulletins. I modernized and streamlined the process of creating the bulletins. Within a few months, I was able to write computer programs to ingest and process the data and to build the tables of data that were needed for both publications. What was taking our data managers dozens of hours per month to complete, I could now finish in a few hours.

My work on the *Solar Indices Bulletin* also put me in charge of the sunspot number forecast, making me an NOAA weather forecaster. Among other things, the SSN forecast was used by spacecraft owners to compute and predict atmospheric drag that slowly changes and degrades satellite orbits. My forecast was used by space weather customers around the globe and was NOAA's only space weather forecast that wasn't produced at SWPC. Today, it is still the primary SSN forecast used by regional warning centers and private industry, but now it is produced by the Solar Influences Data Center (SIDC) in Belgium.

Over the following years, the remaining senior data managers and scientists retired. With each retirement, I was tasked with taking over responsibility for their products. My federal managers at NOAA insisted that the hiring freeze was preventing those positions from being backfilled, but I later found out that was not true. There was a hiring freeze, but exceptions were allowed for inherently federal positions like the one I was serving in. They just wouldn't make an exception for me.

I focused on modernizing the most widely used data products, while shelving production of the others. My division chief, Dr. Bill Denig, told me this was a productive exercise that would encourage users to let us know if they still need the products. If our customers didn't complain that their products weren't available, we were to assume they were no longer important, and we simply stopped the service. At the time, I didn't know my federal supervisors were ignoring established procedures for getting rid of NOAA products, including the need to give the public time to comment.

I also inherited responsibilities for much of our historical data holdings, which were stored floor-to-ceiling, filling a roughly two hundred thousand cubic foot warehouse in Boulder. It was one of the world's greatest treasure troves of historical climate data records (CDR), waiting to be made available to the global scientific community. Similar warehouses were maintained by the National Climatic Data Center and the National Oceanographic Data Center. Collectively, these were the bulk of America's historical climate data archive.

By 2008 I was fully performing nearly all of the functions of a WDC director, but as a contractor. Because this was an inherently federal function, I had to do the work while my division chief carried the formal title. To Bill's credit, he was a dedicated public servant who worked hard and provided me as much support as he could. Bill personally picked up much of the slack that was left when our senior federal data managers retired.

The job kept me pretty busy, and in 2013, I was recognized, along with federal employees Karen Horan and Craig Clark, for "Outstanding Support of the NGDC mission by managing Solar and Geomagnetic databases" by Mary Kicza, the AA for NESDIS.

While I was performing my federal roles, NOAA began to quietly wind down their WDC participation. One consequence of that was I had to continue to perform all of my WDC-A duties, but my appointment was dropped to only 50 percent, and I was forced to take on other obligations. Some of that work was quite fun, but the workload became overwhelming.

In these new roles I was able to support scientific rocket launch campaigns for the National Aeronautics and Space Administration (NASA) and the military. Our world-class ionosonde that operated at the NASA Wallops Flight Facility provided environmental data and science support for those missions. I also had the opportunity to build ionosonde Doppler radar systems all over the world.

Despite the demands of my ever-expanding responsibilities, data stewardship remained my top priority. For several years, I acted as one of the program directors for the Climate Data Modernization Program (CDMP), which was tasked with digitization of historical climate data records. During that program, my team was able to rescue only 15–20 percent of our ionosonde and magnetometer climate data holdings before the program was defunded by Congress.

During these years, it was apparent that there was something causing me to be given such a tremendous workload without being given the opportunity to move into the federal ranks. I trusted Eric that this was caused by funding and hiring limitations that came from Washington DC, and it was only a matter of time before it would be rectified. But there was also clearly an issue of the Marine Geology and Geophysics (MGG) division being favored by NOAA management, and within the NGDC organization, there was a lack of representation for our division in Solar-Terrestrial Physics (STP). I thought that if I could leverage my goodwill with NGDC Director Dr. Chris Fox, I might be able to convince him to mentor Eric as his eventual replacement, and Eric would help correct those internal issues. Ultimately, that became the biggest mistake of my career.

In the following years, I continued to serve NOAA with distinction as WDC-A program leader. But as time wore on, my mistake of ignoring internal NOAA politics caught up with me. Over time, the work situation became increasingly hostile, and I had to back off from pushing Eric to come through on his promise to open a federal position for me.

As a result of years of outstanding work done by Eric's subordinates including myself, and possibly because of my strong advocacy for him, Eric was promoted to acting director. When that happened, his promise to me became a liability.

Unlike our previous director, Eric almost never provided updates to the NGDC team on what was going on within the organization. The organization was undergoing a major transition away from its congressionally mandated functions, and our future was uncertain. Morale collapsed across the center, with our NOAA scientists and data managers no longer knowing what their mission was. Eric quickly became despised by his subordinates as an ineffective leader.

The liability that I created for Eric would become intolerable to him. Eventually, he would direct a colleague to threaten to fire me, strip me of my title as WDC-A program leader, damage my reputation among my NOAA colleagues, blacklist me from federal job openings at other NOAA agencies, and ultimately order the Cooperative Institute to fire me.

CHAPTER 5

THE NATIONAL GEOPHYSICAL DATA DELETION CENTER

When I was hired, Eric explained the Cooperative Institute (CI) was a department within the university that would employ me to work for him. At the CI, I would be a member of the University of Colorado (CU) faculty, my office would be at NOAA, and I would be a NOAA scientist.

The Cooperative Institute for Research in the Environmental Sciences (CIRES) is itself a wholly unremarkable institution. CIRES occupied a floor in a university building, and the office spaces were occupied by administrators. Functionally, CIRES was more or less like a typical academic department. What wasn't obvious on the surface was that the organization operated like a shell company that facilitated hundreds of independent scientific research programs. The CI is a conduit between senior scientists, global policymakers, private industry, governments, and academia. Effectively, the CI barters power and influence, with almost no actual infrastructure "under the hood."

Collection of real-time weather data is a core function of NOAA that is an inherently governmental function. Our program was unique because it was a real-time weather data program that was allowed to operate within the CI, as if it were a typical scientific research program. Fundamentally, it was against the law for NOAA to allow that to happen.

That data program ingests, archives, and disseminates data from ionosondes. The ionosonde was the world's first radar and began making environmental observations in 1925. To this day it remains a simple, robust, and inexpensive radar system[21] and is the most reliable way to make observations of the atmosphere at very high altitudes. Prior to World War II, the earliest continuously operating radar systems were placed in strategic locations to support radio communications and signal intelligence.

One of our longest-running radar data records is the Washington, DC–area ionosonde observations. The current system is located at the NASA Wallops Flight Facility (WFF) rocket launch and flight facility on the Delmarva Peninsula in Virginia. It continues to extend one of our longest, nearly continuously operating environmental radar data records, but it only continues to operate thanks to the personal efforts of a few dedicated scientists.

The Vertical Incidence Pulsed Ionospheric Radar (VIPIR) is currently the world's most advanced ionosonde and was first deployed in 2007 at WFF. The VIPIR ionosonde was the result of years of effort that was the brainchild of Dr. Terrence "Terry" Bullett, in collaboration with the US Air Force, NASA, and a small private business. They developed a programmable ionosonde that could be used as an advanced scientific research instrument and make regular weather observations at the same time. As the world's most advanced ionosonde, it had capabilities I would later leverage to complete my doctoral dissertation.

The WFF VIPIR was deployed alongside an older ionosonde that had been providing real-time weather observations that we consider to be our primary climate data record. To maintain a good baseline for climate observations, we operate the new generation instrument alongside its predecessor. The two ionosondes require regular maintenance, and most years, Terry and I would make a maintenance trip to Wallops.

Our primary mission, and the means by which we supported the NOAA mission, was to manage the global ionosonde radar program for the United States. Terry and I were responsible for quite a lot: We built and maintained field sites, we developed advanced technology for the radar systems, we managed and maintained the international program to collect real-time data from all over the world, we managed real-time systems to deliver that ionosonde data to drive weather models, we conducted scientific research, we provided expert-level support to our partners around the world and throughout our government, we were stewards of our nation's climate data records and international relationships, and we built and maintained radar systems in remote parts of the world.

Because of all of those responsibilities, Terry and I are among America's leading experts in the original radar technology.

Destruction of Climate Data Records

For most environmental scientists, the end of the year is the time to relax. School gets out for those of us who teach, the offices clear out over the Christmas and New Year's holidays, and the year is capped off with the annual American Geophysical Union meeting (AGU). AGU is one of the world's largest social gatherings that masquerades as a scientific conference. Anybody who is anyone in the geophysical sciences is there. If you're not there, you're not square.

A lot happens at AGU. Those of us who are mission-focused might have a beer with colleagues to plan for upcoming campaigns. Another important use of our time is to attend presentations and poster sessions to get up to speed on current research. Similar to what I imagine happens at private meetings of the global elite, the nepotists gather to plot how to run the scientific world.

For me, December has typically been the busiest time of year, and I usually can't attend AGU. The main reason was our government-sponsored project was renewed at the end of each calendar year. It's typical in the federal bureaucracy for funding to be "use or lose." All year long we would be as frugal as we could be and avoid buying hardware until December. Every year we were most busy during the holidays, always scrambling to tie up loose ends and make the most of every dollar the taxpayers provided.

In the fall of 2015 I was sent to South Korea to build two new field sites that were commissioned by the Korean Space Weather Center (KSWC). KSWC is the South Korean equivalent of the NOAA Space Weather Prediction Center (SWPC), one of many weather forecasting centers known as a Regional Warning Center (RWC). One site was located at the KSWC on Jeju Island, and the other site was at Icheon on the Korean Peninsula. I had to manage all of the challenges of doing construction in a foreign country, including working through complex cultural differences. The trip was extraordinarily difficult, and by the time it was finished, I really needed some time off work. But before I flew home from Korea, my plans had been ruined.

I was told that NOAA was hiring three new computer technicians who would start work on January 1. To my surprise, management seemed to think we had no office space left at the sprawling David Skaggs Research Center (DSRC) in Boulder, Colorado. In typical fashion, my federal supervisors lost their collective minds and demanded that we immediately clear out a large data storage room. True to form, they had no plan or intention to find a new place to move the data to, they didn't care that cleaning out a data room was not my responsibility, they didn't care that my labor had to be stolen from other projects to perform the task, and they had no sympathy for the fact that I was coming home from a very hard, two-month-long overseas trip. It had to be done that day. It was an EMERGENCY!

Out of nearly one hundred scientists that worked at the data center, management knew which of us to ask. Only a handful of us cared enough about our climate data records that we would drop what we were doing to prevent data from being tossed in the trash heap. The sad truth is that, although most scientists are honest and hardworking, only a few of us are truly passionate about what we do, believe in the mission, and prioritize our efforts for the greater good. Management knew exactly whom they could take advantage of. Our supervisors leveraged our devotion, and they lied to us to get their unethical and illegal mission accomplished.

Obviously, I couldn't start on it that day because I was thousands of miles away in South Korea. The work had to wait until I got home, and I needed a vacation. I pleaded with management to give us time, but they insisted it could not wait. I canceled my hard-earned and well-deserved family vacation to deal with this fabricated emergency. We accomplished the task, and the freshly emptied and cleaned room was ready for the new employees to arrive on January 1.

For most of the following year, the room was left empty. There were no employees that needed that office space on January 1. Once again, we were deceived by NOAA leadership. Eventually, the room was converted to a garbage dump for obsolete computer hardware the information technology (IT) staff had hoarded over the years. Not a single employee ever took up an office in the room, but the damage that was done to our climate data records (CDRs) can never be undone.

Data was lost. More precisely, data was intentionally destroyed. Even more precisely, senior NOAA federal supervisors had ordered us to destroy CDR's.

I couldn't understand why this was such an emergency. Maybe they really thought they needed the room, or maybe this was part of NOAA leadership's plan to get rid of historical data holdings. What was clear is they knew it was wrong and did it anyway. But as time went on, it would become clear that this was part of a larger conspiracy to get rid of our historical climate data.

Our CDR's are a priceless and irreplaceable treasure. NOAA leadership is very proud of their multi-billion-dollar satellites, which are great for making weather observations but do little to help our understanding of climate change. But, in almost two decades, I never witnessed a senior federal employee make any serious effort, or allow funding to be used, to preserve our irreplaceable CDR's. To the contrary, senior officials would go to great lengths to cut efficient, low-cost data preservation programs that only made up a fraction of a percent of their budgets.

The alleged prejudice was against the so-called unfunded mandate. These unfunded mandates were viewed as things we had to do, but Congress never had the wherewithal to provide the funding we need to do it. In reality, Congress would fund the work, senior leadership would redirect the funds for their pet projects, and we'd have to beg, borrow, and steal the resources we needed to do our jobs. It felt like preservation of climate data, and God forbid, digitization of our scientific climate data records was an unspeakable burden on their billion-dollar satellite budget.

Over almost two decades at the NOAA National Geophysical Data Center, I witnessed only one real effort to preserve our climate record. That project originated as a pork-barrel line item in the congressional budget. The Climate Data Modernization Program (CDMP) was sponsored by West Virginia senator Robert Byrd and was intended to employ ex-convicts to convert historical at risk CDR's into digital data records that could be used by the scientific community. The CDMP effort was the most excellent use of taxpayer money I had witnessed throughout my government career. The project was swiftly removed from the government budget after Senator Byrd passed away.

During the CDMP, I was the project leader for rescue of ionosonde and magnetometer data. Our team of subject-matter experts carefully studied the features of the data records and determined the most efficient and cost-effective way to capture all useful information. We knew that the original data records would never again be processed, and it was incumbent on us to make sure every aspect of our CDR's was preserved during the transition from paper and film to the digital computer archive. This was not a trivial task because most of the data was collected decades ago, and much of the technical expertise needed to fully understand the data had been long lost.

It was also a challenge to work with the contractor to develop an efficient and cost-effective process. I harassed them relentlessly until we managed to bring down the cost from a dollar per page to ten cents per page. Because of our dedicated efforts, we rescued ten times as much data, and we were able to preserve scientifically important information that would have otherwise been lost.

Our magnetometer data records were on 35mm film that was created from the older paper records during an earlier data preservation effort from the 1980s. NGDC was a World Data Center that was created to collaborate with the Russians, and our archive volume was full of data collected in the Soviet Union. To prioritize rescue of the most valuable data, I worked closely with Dr. Anatoly Soloviev, a good friend of mine, who was my counterpart as program leader at the Russian WDC-B for Geomagnetism.

Anatoly was a competent and dedicated scientist who developed an advanced artificial intelligence (AI) algorithm that he used to process the data after it was rescued by the CDMP project. Our collaboration was a triumph of a half century of partnership through the WDC. Anatoly is now the director of the Geophysical Center of the Russian Academy of Sciences. The appointment of a dedicated and competent scientist is evidence that the Russian Federation is still focused on the mission. Together, Anatoly and I identified the best quality and most important data to focus on, and we gave the American taxpayer the best possible value for their money.

For the ionosonde records, my colleagues at NGDC agreed that we should prioritize the ionosonde data ourselves. The decision was to focus on the longest continuously operating stations to establish a more comprehensive climate record. The stations considered top priority included Washington, DC; Maui, Hawaii; and Christchurch, New Zealand.

We managed to rescue 15 to 20 percent of the most important ionosonde and magnetometer data records before the program was defunded. That data can now be discovered and downloaded by anyone in the world, through the NOAA public data archives. The data we weren't able to rescue has since been boxed up and moved to a basement, below the Asheville, North Carolina, federal courthouse. As far as I know, there's no way for a US citizen to research what is supposed to be there in storage. The original inventories were kept in old paper file systems that resembled the Dewey Decimal System card catalogs of old public libraries. I doubt NOAA leadership would support an unfunded mandate to modernize those inventories and put them online for public discovery.

That data, stored in the basement of the federal courthouse, is at risk of being lost forever. I have no doubt that some of the data is lost with every passing day. My predecessors warned me of the many ways that data has been lost.

Loss of Climate Data Records

Stewardship of our historical CDR's is important because they underpin everything we know and understand about climate change comes from those data records. Every theory, model, and forecast was developed using the information contained within this data. Every pattern and trend we've come to expect is understood through the lens of historical observations, and everything we might discover about our past climate lies within them. Future discoveries will come from observations we have not yet made, informed by the historical records we already have.

Simply everything we know about weather and climate comes from our historical climate data record. Without it, we would have no frame of reference for understanding how our climate is changing. With every observation that is lost or destroyed, the breadth of human knowledge decreases.

In the years since the CDMP project ended, most of the remaining experts have retired or passed away. The loss of institutional knowledge means that some of the more obscure features in our historical data records can never be interpreted. Consequently, some of our climate data has been lost forever.

Most people believe that climate change research and climate models have made the best use of all of the data that is available. The truth is that most climate science uses only a small fraction of the available data. In place of data, we apply theory and fudge factors. The idea is that we do the best we can with the resources we have, and future scientists can always go back and improve on our results.

But without a complete climate data record, we can't ever go back and fix our mistakes. The historical climate data record contains information about climate change we have not yet realized. When we lose data, we lose some of our ability to understand climate change forever.

Trying to preserve data is like swimming against the current. Almost every scientist is limited by the resources at their disposal. Assimilating more data into a model, or expanding the scope of a research project, stresses already limited resources. Climate scientists are not pressing for historical climate data to be modernized—they already have their hands full, and there is a reluctance to create more work. The people who profit from the climate change narrative insist that the science is settled and the data can only complicate that narrative. For government, preserving old data records is expensive, and few policy makers understand how important it is.

With little interest in preserving historical climate data, the risk of losing it is severe. As a WDC program leader, my duty was to fight hard to protect this data.

You may have seen old scientific diagrams from the nineteenth century, or even watched a movie with Nicola Tesla standing next to some crazy electromagnetic contraption. Early industrial-era science developed rapidly, with our best minds pushing the limits of technology. Early weather observing instruments were unique, and the data they collected was often inconsistent from one instrument to the next. That made data analysis more difficult, but instruments did evolve to become more standardized over the years.

By the end of World War II, weather monitoring technology was highly advanced, and the network of global weather observations was robust. Coming out of a world war, every country and research group had their own methods and technology, and much of the data was collected by militaries where it was affected by complex security rules and the strong personalities of officers. There was a growing need to standardize data-collection methods and how data was formatted. There was also a serious problem of data records being destroyed when small science programs were dissolved, especially in underdeveloped parts of the world.

These and other factors such as mismanagement, bureaucracy, lack of foresight, cost-cutting, and even deliberate sabotage have resulted in the loss of our climate data records.

Over the years, our vast, comprehensive climate data record ended up in storage at the WDCs. That is where most of it remained for more than a half century. Over that time, quite a lot of data was lost, and much will still be lost. Modernization, through digitization, is the only practical way to save it.

For now, much of the climate data that NOAA is mandated to steward is out of reach of the scientific community. A few years ago, Anatoly and I were discussing how mismanagement was putting our data at risk, and he was sympathetic to that very difficult problem. In Russia he had a similar problem that was giving him a headache. His Cold War–era data managers were mostly elderly Russian babushkas[ii] who were not shy about voicing their concerns. Frankly, they were a real pain in his butt, but he understood why they were so passionate about their work, and he worked hard to try to accommodate them.

---

ii Babushka is "grandmother" in Russian. Here it refers to the matriarchs of the Geophysical Institute of the Russian Academy of Sciences.

At first, he had them bring one box from storage, to the office, and let them digitize the data. Then another box, and another, bit by bit. Eventually, Anatoly convinced the Russian government to allow him to hire a half dozen full-time employees to work exclusively on climate data modernization. Despite having limited resources, the Russians are doing more to protect our climate record than the supposedly climate-conscious US government. I can't help but assume that it would have been easier for them to do what the Americans had done and just bury all the old data beneath the Kremlin wall.

At NGDC, our Cold War–era data managers were just as protective of our data as the Russian data managers were of theirs. Before my WDC mentors retired, they told me the stories of how data has been lost and reminded me that I was all that stood between the data and its destruction. They worked feverishly during the final days of their careers to transfer their knowledge and passion to me, so there would be some hope that their legacy would persist, and our climate data records would be preserved.

The data managers I worked with are all well known in the global scientific community, and they all deserve our gratitude and our respect.

The data manager that I viewed as the most effective bureaucrat was Ron Buhman. Ron was an outstanding civil servant who understood how federal management behaved, and leveraged that skill to create an efficient work environment. He called the massive NGDC historical data archive the "tar-baby," which was a euphemism for a data archive that nobody wanted. It was well known that dealing with the tar-baby only ever created new problems.

The federal scientist who preceded me as program leader for WDC-A for Geomagnetism was Sue McLean. Sue was also very protective of the data, but she was younger and had ambitions to move up the federal chain of command. She was a good fit for leadership because she did care about the mission, but she also had the demeanor of an effective federal supervisor. Her biggest flaw was an apparent dislike for men, which probably came from her experiences in the patriarchal environment she built her career within. Sue surrounded herself with female subordinates, often at the expense of more capable scientists. She seemed to be constrained when sharing stories of how data had been lost and how we need to protect it, but she did tell me what she thought I needed to know. Sue passed on the knowledge I needed to preserve her legacy of protecting geomagnetic data.

In one of the stories, the roof blew off of a data shelter, and a rain storm destroyed many boxes of paper data records before anyone noticed. Her team spent weeks carefully pulling out each soaked sheet and hanging it up to air dry. That was a warning to me to not be complacent that our data is stored in a seemingly secure warehouse. That story proved to be important years later when a damaged roof began leaking and soaked boxes of original ionosonde data records that were nearly a century old.

The most significant data loss was the total destruction of the core of the geomagnetic data program. The United States was an early pioneer in geomagnetic observations, and we had collected an extensive climate archive dating back to the mid-nineteenth century. Geomagnetism was a core program of the Coast and Geodetic Survey, which is one of the three organizations that eventually became NOAA.

Nearly the entire set of North American data was recorded on paper rolls and carefully cataloged and tracked so it could be accessed by the scientific community. But the unfunded mandate, which makes renting storage space cost-prohibitive for the federal government, resulted in a catastrophic administrative error.

The entire data collection was sent to the Denver Federal Records Center for "temporary" storage, but in the federal government, few things are temporary when they accomplish the goal of removing a problem from plain sight. Because of what seems to be a clerical error, the climate data records were misclassified as a low-priority government record, akin to an unimportant email chain. After five years, some bureaucrat pulled the data up for review, concluded some arbitrary hold time had expired, and threw the entire archive in the trash.

I was later told that a disgruntled data center employee was transferred to the federal records center where they intentionally caused the records to be destroyed in an act of vengeance. It's hard to believe a civil servant could be so malicious, but conditions in the federal bureaucracy can foster vengeful and sadistic behavior.

Our STP division chief was Bill Denig, who was a dedicated public servant. Bill tended to blurt out whatever was on his mind, which probably slowed down his career progress, but he was always dedicated to the mission. Bill had come from the Air Force Research Laboratory, and like a typical Department of Defense employee, his idea of the mission seemed to always be what his superiors said it was. Bill was a real scientist, and he understood the value of our data, but he did not view protecting our climate data record as a primary mission. Bill's focus on the needs of his superiors seemed to serve him well. On multiple occasions, that helped him recognize imminent, negative organizational changes before anyone else.

Bill was heavily engaged in a wide scope of spacecraft observations and operations, including data collection and archive, algorithm development for data interpretation, and hardware issues caused by spacecraft anomalies such as the effects of micrometeor impacts. Historical data was not high up on Bill's agenda, and allocating resources for historical data management was not one of his priorities. Still, he did know how bad the situation was, and along with devoting many hours of his time to help us, he would sometimes show his concern.

One major data loss was the comprehensive suite of data collected by our first generation Geostationary Operational Environmental Satellites (GOES). Few people know that all of the data from the first half-dozen GOES satellites had been lost by NOAA. Actually, the data itself was not lost but stored in the NGDC digital archives, which had been well maintained. But a data manager at the Space Environment Center (the predecessor of SWPC) had retired. For many career government employees, including myself, the way that we leave our office on our final day is the way that it remains until someone comes along and throws all that is left in the trash bin.

The GOES data was stored in a binary format, which required an encryption key to be interpreted. As the story was told, the data manager left the only known copy of that key on the shelf of his office. It probably sat on the shelf for six months before somebody hauled an oversized trash can in and threw it away. No matter how this happened, it was lost, and with it, billions of dollars' worth of our oldest and irreplaceable satellite data.

Maybe the GOES data can be recovered by a talented code breaker, but my experience suggests that storing seemingly useless data in a digital archive is considered an unacceptable unfunded mandate. Eventually that data will be destroyed, if it has not already been. Most likely this will be in the next few years because NOAA has mandated that all mission computer systems be replaced by expensive cloud-based services. The bureaucrat who decides what we pay Google and Amazon outrageous fees to store for us will undoubtedly use the transition as an opportunity to write it off for good.

In my personal experience, data is lost in many unpredictable ways. There was one particular way in which quite a lot of data had deteriorated to the point where it was almost not recoverable, but with some care I was able to rescue a good portion of it.

In the early days of radar, the ionosonde observations were projected onto light tables in a dark room. An operator would scale the projected images by hand and fill in paper forms to record the data values. I was responsible for approximately one thousand station years of this paper data, which was stored in more than a thousand boxes. The stuff was nasty. Most of the boxes had not been opened in decades and the stacks of paper, much of it from the 1940s and 1950s were chock full of chemicals. A good lungful of debris from an opened box could send me on a fifteen-minute coughing fit.

Aside from challenges we faced interpreting the data, it was mostly well preserved. But during and following World War II, ionosonde data was considered a military secret. Many of the paper forms were still marked "classified" or "top secret," sometimes in a foreign language. For example, the United States seized all of the Japanese military assets at the end of the war, and eventually I became steward of the ionosonde data from the Japanese Imperial Navy.

Being a military asset, the data collection was supervised by military officers, many of whom were anal-retentive or otherwise demanded that things be done in their peculiar way. For a several-year period, the Washington, DC, ionosonde data collection was managed by an officer who insisted that the information at the top header of each data sheet could not be handwritten. That included important information like date, time, and data type. The large sheets wouldn't fit into a common typewriter, so he had the operators type out the information on separate pieces of paper and cut and paste them onto the data sheets. I can only guess he tried to have them do this with every data entry before realizing that was too much work.

This data had been stored in the same boxes for a half century, and when I started taking it out, I had little bits of paper flaking off. The glue that was used had deteriorated, and the critical information was falling off the data sheets. Luckily, only a few sheets had been damaged before I realized what was going on. That began a laborious process of carefully removing the paper stacks from the boxes and securing each new page as I took away the one that was above it, sandwiching the metadata to its respective data record.

There is also the Good Samaritan who loses data. Our dedicated data stewards do their jobs so the scientific community can use the data for research, to advance the knowledge of mankind. Sometimes this meant sending unique data volumes to scientific researchers for "short" periods of time. Those volumes didn't always make their way back to the data center and are now lost.

After the entire set of North American geomagnetic observations was destroyed, the NOAA historical geomagnetic library was all that was left of many important observations made during major solar storms. Scientists would borrow the most interesting volumes for their research, and they sometimes forgot to return them. After decades of accidental loss, the library is missing the books that contained the most interesting data.

There are many ways we know data has been lost and an infinite number of ways we haven't yet even thought of. Data stewardship relies on a lot of guesswork and has only been effective when it is managed by people who truly care about our climate data records. Supervisors losing their collective minds over a few office seats they don't really need and bureaucrats searching for every way to purge the government of unfunded mandates create some of the worst and riskiest situations. By its nature, the United States government is a terrible steward of important records.

CHAPTER 6

HIDING DATA FROM THE PUBLIC

Old media that is nondigital, or analog, deteriorates over time. One of the most significant losses of American culture happened while storing celluloid film produced in the early days of Hollywood cinema. In addition to being quite flammable, the film would naturally deteriorate over the years. When sound and eventually color were added to movies, Hollywood diverted attention away from these early works of art, and they were all but forgotten while in storage. In storage they began to deteriorate.

Hollywood's efforts to protect and restore the remaining cinematic treasures is ongoing.[22] Sadly, much of these great works of art, along with an important piece of our history and culture, were lost forever.

Climate data records also degrade over time, and glued-on headers, drying up and falling off, are only one way that happens. Over the years, NOAA's dedicated data stewards had identified at-risk and high-value datasets and separated them from the large dry-storage warehouse. Those records were brought into the data center's main office building where some rooms were outfitted with temperature and humidity control systems. These climate-controlled rooms were designed to limit the degradation of our most important data records so they could be preserved until somebody could find the means to modernize the media.

The climate-controlled room that NOAA leadership was determined to convert to a computer hardware waste heap held several high-value datasets. This included the geomagnetic library, which was the most comprehensive library of historical geomagnetic observations that could be found anywhere in the world. Many volumes date back to the mid-nineteenth century, and these nearly two-hundred-year-old books are so fragile that bare fingers can strip the ink from pages instantly and pages could disintegrate just from trying to open the book. A result of steadfast nagging, I was able to convince my supervisors to give me a small climate-controlled room to move the library to. Fortunately, we had one such room, more accurately described as a closet, that wasn't large enough to be converted to office space. Before I was fired, I had made tremendous progress to have the library relocated to the United States Geological Survey (USGS) geomagnetic program offices for proper, long-term stewardship. At USGS, the competent leadership team would have been a far better steward of the library, and they would be subject to the watchful eye of Dr. Jeffrey Love, the library's most frequent patron. My unexpected departure from NOAA could have sunk that effort and put the library at risk, but Dr. Love has since taken the initiative to protect the library.

The other datasets that were stored in the climate-controlled room included handmade plots of derived ionospheric parameters called F-plots. Those F-plots were an important tool to understand daily and seasonal variations that occur in the atmosphere and the related understanding of how climate change is manifested at high altitudes. The volume of F-plots represented thousands of hours of work done by experts who are no longer with us, using a skill set that has all but been lost. In theory, they could be reproduced from the master set of ionosonde data, but realistically that will not ever happen. Because these were a derived product, NOAA supervisors declared they were not data and ordered us to throw them in the trash.

As a WDC, we held quite a lot of data that could be found at a partner data center in another country. For geomagnetic data, the partner data center was WDC-C for Geomagnetism in Kyoto, Japan. Our Japanese partners are known to be thorough, and they maintain a very nice web-based data portal. The International Council for Scientific Unions (ICSU), the UN agency that oversaw the WDCs, wanted multiple WDCs to have copies of every dataset, so there would be some redundancy. But my federal supervisors decided that because the UN dissolved the WDCs, it was okay to throw away data if we could verify it was stored at another data center. I was ordered to throw away all of the magnetometer data that was listed as being in the Japanese inventory.

Dr. Arnaud Chulliat had recently come to NOAA from Institut de Physique du Globe de Paris (IPG), to replace Dr. Stefan Maus as the head of the NOAA flagship World Magnetic Model program. I had met Arnaud several years earlier at a meeting in Uglich, Russia, and we got along very well. Arnaud was intelligent and seemed to genuinely care about the data. In fact, he was one of the only other NGDC scientists who would be willing to take the time to help preserve data. Knowing that, I asked him to help with our "triage"[iii] of the magnetometer data.

One data set was carefully filed on a large rack of bookshelves. The data was from observation stations that most people in our community had never heard of, many from countries in Africa and other remote parts of the world. Much of this data represents the only magnetic field observations ever made near their locations. Most of the data consisted of printouts from 1980s-era printers, and dealing with them was not very easy or pleasant. Arnaud and I methodically sorted through every folder and cross-referenced the contents with the Kyoto metadata record. Almost all of the data was showing as being stored at the Kyoto data center, so we followed our marching orders and began to fill a large trash bin with it.

I vaguely remembered Sue warning me that the dataset was unique and important when I took over as program leader for WDC-A for Geomagnetism. Something didn't feel right, and I told Arnaud that we had to stop and confirm that the Kyoto WDC really did have a copy of all of the data. Arnaud agreed, and he went back to his office to investigate while I continued executing my orders to throw everything else away.

---

iii Our use of "triage" described the rush to prioritize data sets and decide what was to be protected and what was to be destroyed, given our time and resource limitations.

What Arnaud found is that the Japanese team generated their metadata file from a combined inventory of all of the WDCs. The Japanese web links were empty, and they did not have any of the data. We had the only copies that existed anywhere in the world. Luckily, we were able to pull the mess from the trash and reshelve it. This cost us a few days of our time, but it was better than losing the data forever.

That data was largely from remote parts of the world like rural Africa. Had the data been thrown away, we would have done the same thing that we did with the data from North America, when the Denver Federal Records Center made their mistake. Collectively, we would have trashed half of the world's geomagnetic climate data record.

I had just enough space left in the climate-controlled closet that I was able to box this data up and put it there for safe keeping. As far as I know, that is where it remains, with no one at NOAA left to identify it as unique and protect it.

A fairly large volume of high-value solar data also took up a chunk of space in the climate-controlled data room. Fortunately, we were right down the road from the National Center for Atmospheric Research High Altitude Observatory (HAO). One of the senior HAO supervisors was Dr. Scott McIntosh, who is also one of the world's most outstanding solar climate scientists. In his science role, Scott's solar cycle forecast consistently beats out NOAA and NASA. Despite billions of dollars of satellite assets being at-stake, NOAA consistently disregards Scott's work because he's not part of the good old boys' club. Scott is a real scientist and role model, and when I told him that our solar climate data was at risk, he sprang to action and had his team at HAO assume stewardship of it. Handing this data over to a better steward is one of the accomplishments that I am most proud of, and just recently I was told that the data is actively being used by scientists.

Most of the other data was piled up on pallets and shipped off to the Denver Federal Records Center, where it probably awaits the same fate as the geomagnetic data. Now, there seems to be no one left at NOAA who even cares, and the risk to that data is exponentially higher.

We did move some data to another climate-controlled room that was managed by the MGG Division. That data included a single copy of each of the publications produced by NOAA, along with some other data sets that management wanted to hold onto. My suspicion is that their fate awaits the next data purge. After the COVID lockdowns, the NOAA scientists who worked at DSRC weren't forced to return to the office, and I've heard the place was like a ghost town for the following years. I'm not sure why contractor scientists need federal government credentials and site access if they never show up to the office, but maybe the almost empty building bought us a few years before the next data purge.

The biggest heartbreak for me personally was the loss of a small but irreplaceable dataset. During the Cold War, the world's nuclear powers conducted thousands of nuclear tests. The US government was conducting above-ground and aerial tests at Johnston Atoll in the Pacific Ocean. They installed an ionosonde to monitor the atmospheric effects of those nuclear explosions. The data they collected was classified for decades and stored at the National Institute for Standards and Technology (NIST) which is collocated with the NOAA facility in Boulder. For years, NGDC scientist Ray Conkright pressured NIST to release the nuclear test data to NOAA. Ray eventually wore them down, and the data was declassified and transferred to NOAA. The data consisted of about a dozen reels of 35mm film, and we quickly targeted it as high priority for the CDMP project. Unfortunately, the CDMP project had ended before the film reels were processed. They were placed in a small cardboard box and left in the climate-controlled room until the triage.

When I was preparing to defend my doctoral dissertation, I found some old papers that described strange features that could be seen in radar data at times when nuclear testing took place. I worked to investigate if the shock waves from nuclear blasts were similar to the shock waves created when rockets are launched, a volcano erupts, or an earthquake takes place. This would have been important work toward comparing natural atmospheric phenomena with their man-made (anthropogenic) counterparts. We had an amazing, never-before-published dataset from nuclear testing, a few doors down from my office, waiting to be analyzed.

But the data was gone. Somehow, during the data room triage, somebody took the unmarked box and tossed it in the trash. With above-ground nuclear testing now banned, that dataset can never be reproduced, and until nuclear Armageddon takes place and I can conduct a fresh study, my research cannot be completed. Terry and I both remember exactly where the box was kept; it was on the floor to the left as soon as you walked into the room. Now it is lost forever.

The Warehouse Move

After the data room triage, I worked with my colleague Dave Fischman to develop a preemptive plan to prevent the same mistakes from being made again. Our entire team was pretty freaked out by management's blatant disregard for the need to preserve climate data records, and we agreed something similar was probably going to happen again. When Dave and I finished our proposal, we submitted it to NOAA leadership for consideration.

Our proposal highlighted the fact that we were leasing a large, expensive warehouse. We told them we were concerned that NOAA leadership would hastily decide to clear out that warehouse and put the data at risk. Our proposed plan was to organize all of the climate data records, create an updated inventory so the mess could be tracked, and to develop a long-term data stewardship plan. The long-term plan included digitizing the data and making it available to the scientific community.

That proposal seemed to have the effect of alerting management to the problem, and true to form, they ignored our concerns and took the most irresponsible action possible. Management decided that all of the data needed to be reboxed so it would be ready to move. In hindsight we know that leadership had already secretly decided to move the data to Asheville, North Carolina, and this was just their way to get ready for that. Until Dave and I raised our concerns, NOAA leadership merely had it on the to-do list, with no plan for the logistics, no concern for protecting and preserving the data, and no understanding of what a responsible move would require. Dave and I unwittingly tipped them off, and our good intentions would be thoroughly punished. Reboxing all of our data had become a new emergency.

We had alerted management that there would be a Herculean effort to reorganize everything in our warehouse before any move could take place. In response, they decided that we had to use standard-size legal boxes that were designed to store typical 8.5-by-11-inch printer paper. These federal supervisors, who were responsible for our nation's climate data records, didn't care that scientific data wasn't usually recorded on standard printer paper but was recorded by hand on oversized and nonstandard formats. Some of the stuff was on relatively manageable eighteen-inch-wide paper, but lots of it was on poster-size paper that was either rolled up or stored in huge cabinets designed to store large-format maps. Our supervisors even insisted that thousands of 35mm film reels be removed from specially designed cabinets and stacked in these boxes for transport. Leadership had ordered exactly what we warned them we had to be careful not to do. The orders were so irresponsible that it was hard to believe they were not intentionally making the data harder to manage. The action was the very definition of fraud, waste, and abuse.

As management pressed forward with their plan, they kept the data managers in the dark so we wouldn't raise concerns. As a WDC-A program leader, I was responsible for protecting the data. When I found out what was going on, I went to NCEI Deputy Director Eric Kihn and insisted that I be included in the planning process. Eric knew there were a bunch of issues with the proposed plan, and he knew that I was well intended and wanted to help management get out ahead of them.

I had already established that our oldest ionosonde data records were fragile and had to be taken out of the boxes with great care, or the planned reboxing by contractors would destroy a bunch of those records instantly. The bulk of the ionosonde data records were stored on nearly ten thousand large film reels that were unwieldy and were each labeled with a hard-to-read four-digit, handwritten number on the side of each tin 35mm film can. Years earlier, two data managers undertook a huge effort to reorganize those reels, after an earlier warehouse relocation had scrambled them up. Management's plan would badly mix up the reels, cost thousands of man hours of time, and ensure those records could not be accessed for years.

Our entire collection of cosmic radio noise data was stored on flimsy paper rolls that resembled old-school cash register paper. I was told to throw the entire archive in the trash, despite the fact that they contain important and irreplaceable information about the Sun's effect on climate change. Fortunately, I convinced Bill they needed to be boxed up with everything else, and they have been spared for now. Spacecraft observations of Jupiter were in oversized hardcover folders, and the metadata that was needed to interpret the data was in a separate pile of hardcover folders. Had those been separated, the data would have become useless. Some old data-reading machines that were needed to read obsolete formats were tossed in the dumpster with no plan on how we would ever read the data. None of the data in the warehouse fit properly into the boxes, and much of the air we were breathing in the warehouse was filled with remnants of data sheets that would become aerosolized as we opened the boxes. We, the subject-matter experts, knew exactly what we were up against, and leadership did everything they could to thwart the effort.

As it turned out, the reboxing effort was actually the start of a plan to move the entire archive. NOAA leadership had long planned to move all of our historic climate data records, but they were keeping it secret from the scientists and data managers.

The federal warehouse manager in Asheville was Jason Cooper. Jason was a former TV meteorologist who decided to change his career path and took a position at NOAA. I really enjoyed working with him, and as far as I can tell, he is a dedicated public servant and was the right person for the very difficult job. When we met, the first thing I told him was that he was walking into a huge mess. I don't think he really understood what I was trying to warn him of, but when I showed up in Asheville a few years later, he confirmed that I was right all along.

Jason had a job to do, and from management's perspective, their pesky subject-matter experts were interfering. There were two lead supervisors on the effort: Kelly Prendergast was a space physicist I had worked with years earlier who had become the division chief of the Information Technology branch that was responsible for our computer systems; Scott Hausman, who had recently come over from the Air Force, was appointed chief of the newly created Data Service Division, which now claimed ownership of all the historical data. What we understood is that Kelly and Scott were following orders from senior leadership.

I reached out to Kelly and Scott to complain that the data managers weren't notified about the upcoming data move. I pointed out that the effort seemed to be a response to the proposal Dave and I submitted, but they were taking every action we warned them against. I told them that this was unsecured climate data that was at risk, and it shouldn't be treated like an already secured data archive. I impressed on them that we were dealing with problems that had been ignored by management for years, and it was a challenging and complicated task. As a WDC-A program leader, it was my duty to help.

Kelly and Scott completely ignored my concerns but initially they seemed to recognize that I was trying to do the right thing. As the process moved forward, they kept pressing for every irresponsible method we had advised against. Eventually I had to tell Jason that if they continued to ignore my concerns that I would have to put a stop to the process.

At this time, I thought Eric and I were still friends. I felt comfortable being direct and honest with him, and I believed he would do the right thing.

Eric called me into his office and asked what I meant when I said I would put a stop to the warehouse move. I explained to him that I was a representative of the WDC and had an obligation to protect the data. If they continued on their path, I would have an obligation to raise the issue with the appropriate authorities and blow the whistle. I told him that I was a team player, I knew that the records had to be moved, and I only wanted to be involved with ensuring they are transferred safely. I told Eric I wanted to be given a heads-up before anything moves, and he promised me he would keep me in the loop.

The connection I failed to make was that Tom Karl's manipulation of the temperature records to influence the Paris Agreement was part of a secret policy at NOAA. Since climate change was "settled," the old policy of preserving our climate data records had been replaced with a new policy to manipulate and hide climate data records. When my colleague John Bates exposed what Tom did with the temperature record, he put the entire scheme at risk, and NOAA leadership was in the process of doing damage control. My threat to blow the whistle on their irresponsible data-handling policies risked blowing the lid off that scheme, and I had to be stopped at all costs.

As the effort progressed, very few of the NOAA scientists and subject-matter experts were helping. Everyone was asked to help, but most of them couldn't care less. Only a few of us had enough concern to find the time, including Terry, Dave, and myself. To get the job done and prevent data loss, we had to drop everything else we were doing over the next couple of months. It was a tough time, but our effort paid off. We helped preserve the data and made Jason's job a little easier. We ended up saving the taxpayer a good portion of the millions of dollars it was going to cost to fix the mess later.

Even Eric would eventually have to admit that much data would have been lost if it were not for our effort. For that, our team was awarded the NOAA bronze medal, a prestigious recognition of outstanding service.

The federal lead on our team was a data clerk who had continuously been passed up for promotion because she dared to file a harassment complaint against a toxic supervisor. The rumor was that the supervisor showed up to her house drunk, expressing his eternal love for her. She was hardworking and resourceful, but she was not a data scientist or a subject-matter expert, and was overwhelmed with the monumental assignment she was given. But she was happy to work in the warehouse and not have to be in the office in a hostile workplace.

That June, she told me that management was pushing to have the entire warehouse moved quickly over the summer. Eric promised me he would keep me in the loop, but he didn't. The next day, I reached out to him and asked to be involved in the process. I told Eric that with Bill Denig's recent retirement, I was the only remaining WDC-A program leader and asked to be recognized as interim director for WDC-A for Solar-Terrestrial Physics. We needed a director for the WDC because it was a collaborative international program that had existed for a half century. There were a tremendous number of loose ends and questions about how to deal with them.

What Eric didn't tell me is that he had already assumed that title and was parading it around to position himself in the international climate change community. He assumed the title to pad his own resume, but he had no intention of doing any of the important work. Eric's point of view seemed to be that the WDC was dissolved and there was no WDC work to do, but the title still came with power and influence that he could leverage. But Eric wasn't interested in what the WDC was and what it still required, and clearly, he didn't care about the consequences. The power and influence were more important to Eric.

The request I made to be part of the process was my duty as a WDC-A program leader, and it shouldn't have caused any problems. But Eric and his boss were betraying the taxpayer and the NOAA mission by moving the data. What I was doing to protect the data risked exposing what they were up to, and they decided that they needed to cover their butts. In what I would find to be Eric's normal way of dealing with problems, he ordered someone else to do the dirty work. He directed my fellow CIRES scientist and colleague Juan Rodriguez, to threaten me.

The Threat

A supervisor ordering their subordinate to do their dirty work is one of the most horrible acts of cowardice that can occur in a workplace. This is especially true in the federal government where the chain of command is hierarchical and people's career progress is heavily dependent on kissing the boss's butt. I'm sure Juan must have felt threatened by Eric, but I also want to believe that he was trying to do the right thing.

The threat came when Terry was on sick leave, recovering from major heart surgery. Eric thought he could leverage Terry's absence by having Juan appoint himself as my acting supervisor. That was news to me because Juan was nothing more to me than a respected colleague and he was less senior in the organization than I was by more than a decade.

They fabricated a chain of command and asserted that because Juan was Terry's Cooperative Institute (CI) supervisor and Terry was my CI supervisor, Juan inherited that role in Terry's absence. The fatal flaw with that logic is that within the CI we were all university faculty. Because of the longstanding, internationally recognized core principles of academic independence, the supervisor role was purely administrative and in no way authoritative.

We were university research scientists who enjoyed all the benefits of academic independence and tenure.[iv] Terry was my actual supervisor because he was the ionosonde program leader and the principal investigator of our

iv Here "tenure" is used in the internationally recognized sense.

projects. In other words, Terry was the senior professor. It was abundantly clear to everyone that Juan and I had no business together.

In Terry's absence, I reported to James Manley, who came to work with us after he retired from the Air Force as a lieutenant colonel. Jim was known and respected throughout our space weather community as one of the legends. I joked with Jim about being the Forest Gump of the Air Force because he was right in the mix of some of our biggest historical events, including the fall of Saigon and the Iranian revolution. When Jim was an officer working with Strategic Missile Command, a solar storm was confused with a Russian nuclear missile launch, and it nearly caused World War III. Jim and his team figured out that it was weather and not a Russian missile launch, saving us all from an imminent nuclear apocalypse.

The first thing Juan did was reply to an email that I sent to Eric. Juan, who himself had no business being involved, included senior CI staff on the email to make sure they knew I was being reprimanded. Juan ordered me, the program leader for WDC-A, to cease and desist all WDC activities. He also claimed that it was inappropriate for me to offer to act as interim director of WDC-A for Solar-Terrestrial Physics. That was a shock given that I was the most senior WDC scientist in the organization and the only person qualified to assume those duties. Juan's email was overtly hostile, and I knew right away that I was being retaliated against for threatening to blow the whistle.

At the time, I wasn't familiar with whistleblower protection laws. When I told Eric that what leadership was doing was wrong, that was a whistleblower complaint. When Eric had Juan attack me, they were both retaliating against and illegally revealing the identity of a whistleblower.

The same day, Eric ordered the IT staff to disable my access to several WDC-A email accounts that I used to communicate with colleagues, customers, and partners around the world. Those email accounts were the primary contact points for three major programs under the WDC-A for Solar-Terrestrial Physics, including my geomagnetism program, the solar physics program where I was the most senior subject matter expert, and the ionospheric observations program, which I was managing in Terry's absence. Nobody told me my access was being disabled; on Eric's orders, they just did it secretly. I noticed almost immediately because I made it a practice to copy myself on WDC communications and it was obvious when my own emails stopped showing up in my email inbox.

I replied to Juan's email in the only way I could and requested that the uninvolved CI employees ignore it. This was a federal data management matter that had nothing to do with the cooperative institute. In fact, it had nothing to do with university business whatsoever, and it wasn't appropriate for Juan to be airing our dirty laundry to uninvolved colleagues. The only reason to do that was to paint me as a troublemaker so he could gain leverage.

One day Juan stopped me in the hall and asked me to come to his office to talk. Before we started, I asked him to explain the email he sent. He said Eric called him into his office in an "official manner," and he told me the conversation he had with Eric wasn't a "water cooler" discussion. Juan tried to relay to me that the conversation was very serious, so it would scare me. Eric told Juan that if he didn't take care of the matter, Eric would have me "thrown out of the building."

Eric's threat was not only a blatant act of whistleblower retaliation but also a cowardly act to ask a low-level subordinate to do his dirty work. At that moment, I forever lost any remaining respect I had for Eric.

I explained to Juan that I had been the program leader for the WDC-A for Geomagnetism for over a decade and I had international obligations that needed to be met. Juan told me that I was not a WDC program leader, so I pulled out my NOAA business card that said otherwise. I told Juan that if my title had been revoked, it would have been prudent to let me know.

But now Juan, in his self-appointed authority, was telling me that I was no longer in that position, claiming that Eric had acted against me. At risk of pointing out the obvious, I let him know that no one in NOAA leadership had told me anything of the sort, and he had no authority to declare it. The basic issue, I explained, is that such an important change had to be done through a process that would protect the integrity of the WDC and my own international reputation. Eric was obligated to work with me if he wanted to remove me from my position.

Juan insisted that Eric working with me was not necessary and continued to insist that he was my supervisor and could direct me how to do my work. I'm pretty sure that as Terry's CI supervisor, Juan never had the guts to tell a senior scientist like Terry how to do his work. It became clear that Juan was handpicked by Eric because he didn't have respect for me.

One research scientist ordering around another scientist is an outrageous assault on their professional integrity. What Juan did to me was way beyond a violation of academic protocol and professional conduct. This was a blatant and shameless act of disrespect.

I calmly let him know that he wasn't my supervisor and I didn't take orders from him. He became visibly irate and started arguing that with Terry on medical leave, he inherited the duties as my supervisor. I respectfully disagreed and told him that I simply did not take orders from him. He became infuriated, his face turned red, he stood up from his desk, and he raised his hand to me and started waving his finger, saying, "Get out! Get out!" It felt like he was about to punch me, which startled me, but I managed to keep my cool. As I sat there calmly, he came to his senses, sat back down, and apologized.

As the conversation continued, the cycle repeated: Juan wouldn't accept that I wasn't going to take orders from him, he became infuriated, and then he threatened me. The surreal and outrageous meeting lasted for more than an hour.

During the meeting, Juan accused me of being a troublemaker with a history of causing trouble. At that time, I had been with the organization for fifteen years and never had any problems until management started breaking laws and putting our climate data at risk. Apparently, doing my duty as a WDC program leader and civil servant is what Eric and Juan saw as being a troublemaker. That accusation was a weak attempt to demean my character while propping himself up on some misplaced moral pedestal.

What was really going on is Deputy Director Kihn sent his subordinate to threaten me, so he wouldn't have to do it himself. Threatening to fire someone for reporting mismanagement is a serious offense, and he might have thought having Juan, do the dirty work would provide some cover. Before that day, I respected Juan and I looked up to him as a role model. Not only was I hurt by what he said to me, but I also would never again feel safe or welcome at work.

As if that wasn't enough, he pressed even further. In a bizarre continuation of the personal attack, Juan claimed that I was not an expert in WDC activities. At that time, I had more experience with our WDC program and holdings than anyone at NOAA, I was by far the most senior WDC-A scientist, and I was recognized internationally as a world-leading WDC expert. Literally and unequivocally, I was America's foremost leading expert for WDC activities in Solar-Terrestrial Physics.

Juan was devoted to his insane, uninformed insult. He claimed that the WDC experts were in Arnaud's team that was responsible for the organization's flagship World Magnetic Model. Arnaud's group worked almost exclusively with satellite data, and cumulatively had very little experience with, or knowledge of, the WDC. Juan's approach was an outrageous position to take for the sole purpose of trying to discredit my expertise.

For context, one very common personality flaw that scientists have is a tendency to insult their peers. Typically, the insulter is a scientist who needs to be the center of attention by making someone else look like an idiot. That sort of behavior is normally tolerated in the scientific community, but I've always felt that it has no place in an academic environment. Unfortunately, that behavior is not likely to be corrected anytime soon because our community is just too small to become the focus of any sort of social justice movement that would shame scientists who were disrespectful of their colleagues.

At the same time, scientists are important, and manipulative behavior can reap huge benefits. Publicly disrespecting a colleague is an egregious practice that harms people's professional images in front of their small peer groups and can have a terrible effect on their reputation. Across all the scientific disciplines that I have witnessed firsthand, the self-righteous communities of high-minded scientists have tolerated this abhorrent behavior.

At least Juan didn't try to humiliate me publicly; he only humiliated me in front of a small group of my colleagues. It was out of character for Juan to attack me the way he did, and I have to believe he was under immense pressure from Eric. Juan was not the kind of person to publicly insult his colleagues, but still, he didn't have to dig too deep to find his dark side.

I did my best to be cordial and quell the situation. I told Juan some of the WDC stories of how data had been lost and how the current effort was putting our climate data records at risk. I even tried to be understanding of his position and not hold it against him, especially because he told me that Eric had issued the orders. But Juan would not budge—he continued to insist that protecting our climate data records was not my duty and that I was causing problems. He didn't seem to recognize the plainly obvious fact that it was NOAA's federal leadership, and not me, who was causing the problems. This type of attitude is common and helps show how and why the federal government is rotten across almost all federal agencies. People are so motivated to suck up to leadership, they can't see when they are betraying the mission and the taxpayer.

What was really frustrating is that I had gone out of my way to work with leadership and avoid causing any problems. I clearly explained to Eric that I would go along with the decision to move our data and that I would help facilitate the move. The only thing I asked for was to be allowed to help with the process and make sure that our climate data records were protected. I had even reiterated my loyalty to Eric and my willingness to do whatever it took to get the job done. The fact that Eric had Juan attack me for that is proof that they were doing something wrong and they knew it.

What I still can't figure out is why Eric went after me, the one guy who had been loyal to him for years. This overtly hostile act showed that Eric was willing to act without conscience and without considering the repercussions of his actions. Why was he so eager to burn the one guy who he knew would do almost anything to protect him? At the time, I could only think of one possible answer: For years Eric had me do his dirty work while serving as a contractor in a federal role. I knew where the metaphorical bodies were buried, and his misguided sense of self-preservation convinced him it was safer to get rid of me than it was to make peace and give me what he had promised.

Juan and I simply had to agree to disagree. He was not my supervisor, and Eric's plot to have him order me around didn't work.

At the end of the meeting, Juan stressed that he was only trying to help, and he told me what he would do if he were in my position. I let him know that I respected his view, but we are individuals and have to make our own decisions. I told him that as a matter of my faith, I am, at any time, compelled to do what I feel is right.[v]

So, within a few days of me making a whistleblower complaint to Eric, he had Juan attack me and make me look like a troublemaker to my colleagues, tried having the university discipline me, and harmed my professional relationships and standing with my colleagues at NOAA and around the world. After years of loyalty and friendship, Eric had betrayed me and launched a full-scale war on my reputation and standing.

When sending Juan to attack me didn't work, Eric turned to senior university leadership.

The Cooperative Institute to the Rescue

The Cooperative Institute for Research in Environmental Sciences (CIRES) is the oldest, largest, and most influential CI throughout NOAA.[23] CIRES is sponsored by NOAA and hosted by the University of Colorado at Boulder (CU). Within the university, CIRES is functionally the equivalent of an academic department that is run by professors.[vi]

The professors at CIRES are research scientists, not classroom teachers. CIRES only poses as an academic department, when in reality it is merely hosted by the university. The CU Board of Regents allows CIRES to act as if it were a completely independent organization. As long as they abide by university administrative rules, CU leadership turns a blind eye to what the CI does, especially inside the walls of the federal workplace.

Today, CIRES takes in hundreds of millions of dollars a year. The CIRES director takes credit for the work of hundreds of NOAA research scientists, with absolutely no accountability to the federal government's chain of command. This is a situation that was designed to facilitate corruption. Similar to how Dr. James Hansen, who posed as a climate change expert merely because he was a NASA laboratory director, the CIRES

---

v My comment was a reference to Thoreau in his essay on the duty of civil disobedience.

vi My use of "professor" includes researchers who don't teach. That is the internationally recognized meaning.

director could use his administrative role to claim expertise over nearly every aspect of NOAA funded research.

After my meeting with Juan, I received a bizarre email from the CI director's deputy, Christine Wiedinmyer. Christine is a senior, internationally recognized climate scientist who even has her own Wiki page.[24] She reasserted Juan's fabricated and complex chain of command, and like Juan, she had inserted herself within it and assumed the authority to give me orders. She concluded by saying "I appreciate your work for CIRES, and I hope that we can continue this productive arrangement."

Simply put, this well-respected, internationally recognized scientist compromised her integrity and risked her reputation by assuming a nonexistent authority to tell me what to do. As with before, the order was to stop me from doing my job of protecting irreplaceable climate data, and she used a veiled threat of termination if I didn't comply.

After Christine's email, I received the first attack from the CI Director Waleed Abdalati. He wrote to me to claim that my actions were not representative of the university's standards. It was surprising for me to learn that firmly holding to my principles with integrity was contrary to what a university expects of its professors. Abdalati proceeded to give me a lecture about how whistleblowing had a very specific meaning in the federal government and that I was a contractor and therefore not allowed to report fraud, waste, and abuse to the government. According to Abdalati, I was only allowed to report federal misconduct to university officials.

Abdalati was obligated to take the same training courses as we were, and he must have known that contractors are supposed to report misconduct directly to the federal government. I wonder if he has the same views for sexual misconduct, racism, or harassment? Telling me that I was not allowed to report fraud, waste, or abuse was wrong and illegal. Threatening me was another retaliatory act against a government whistleblower.

Filing a formal complaint to the Inspector General (IG) was a hard decision. Retaliation is taken against almost all whistleblowers, and I was sure that these psychopaths would hold a grudge. After talking it over with my wife, we agreed that they were determined to get rid of me and the IG complaint was the only way to protect myself.

The IG complaint did stop the overt attacks, but only for a few years. They should have all been punished for their actions, but the IG improperly classified my complaint as an initial whistleblower complaint. It should have been classified as a report of whistleblower retaliation. As a result, the IG sent the complaint back to NOAA to investigate themselves, instead of launching the required IG-led whistleblower retaliation investigation. As expected, NOAA buried the complaint and concluded that my concerns of retaliation were unfounded.

For the next several years, Eric Kihn and Waleed Abdalati worked quietly to limit my success, ensure I was uncomfortable in the office, and set me up so they could eventually find an excuse to get rid of me. They knew that I was imperfect and if they were patient and kept up the pressure, they would eventually find something they could use to "throw me out of the building."

CHAPTER 7

VOLUNTARY NARCISSISTIC RAGE

Waleed Abdalati is a handsome, bald man of Middle Eastern descent. He was visibly muscular and very charismatic. As CIRES director, Abdalati would host our annual CIRES Rendezvous that most of their nearly one thousand scientists would attend. In his first couple of years as director, Abdalati would break into a rendition of Frank Sinatra and dazzle the audience with his charm and talented voice. The ladies would swoon, and their request for him to sing became an annual favorite of the "secret ballot."

Abdalati was the graduate student of the previous CIRES director, Konrad "Koni" Steffan. Koni left CIRES in 2012 at the request of the president of the Swiss Confederation to become director of the Swiss Federal Institute for Forest, Snow and Landscape Research. In August 2020 Koni died when he fell into a crevasse on a Greenland glacier. I sent Waleed my condolences and told him, "There's a place in Valhalla for scientists who die with their boots on."[vii]

Years earlier, Abdalati was appointed by President Barack Obama to be the chief scientist for NASA. A large part of NASA's transition away from space flight and into climate change happened during Abdalati's watch. Some of that work was undone during President Donald Trump's

vii I'm paraphrasing here because I don't have access to my NOAA email.

administration, which gave me hope that NASA might return to its prominence as the world-leading organization for space exploration. It was a disappointment, but no surprise that President-Elect Joseph Biden put Abdalati on his transition team to help undo changes made by the Trump administration.

The question I can't answer is why didn't Abdalati take a senior position in the Biden administration? Instead of seizing on that opportunity or taking advantage of the DC swamp's revolving door to enrich himself, he decided to remain as director of the CI. We will explore his role in the global climate change industry and whom he might answer to. Possible answers to that question will start to take form.

A Russian Agent

Dr. Nikolay Zabotin is a Russian expatriate whom I have never witnessed smoking, drinking, or cursing. I viewed Nikolay as an excellent scientist I wanted to collaborate with, but he had been creating problems for our research program for years. Like Terry and me, he is an expert in signals intelligence, specializing in interpretation of radio wave propagation physics.

Terry and Nikolay had worked closely together for decades. Nikolay tried to assert himself as the senior scientist and group leader of Terry's program, consistent with the hierarchical structure found in a Russian academic environment. Terry would push back, maintaining an independence between his instrument program and Nikolay's research. This frustrated Nikokay, leading to a condescending attitude toward Terry, furthering their division. Eventually, Terry decided to pare back his interactions with Nikolay, limiting his access to what was needed to accomplish our mission. In response, Nikolay resorted to devious methods to obtain privileged information and access to hardware and facilities. On one occasion, he tried to gain unauthorized access to a NASA base, and on another, he tried to obtain details on an Air Force research experiment. He often acted suspiciously, so we would worry that he might be up to something any time he was pretending to act "normal." Because of his personality, we affectionately gave him the nickname "KGB."[viii]

viii KGB refers to the Soviet Committee for State Security.

Most of the problems that Nikolay created were tolerable. When Terry and I exchanged scientific ideas with him, it was common for him to dismiss our ideas as being terrible and then leverage those same "terrible" ideas for his own benefit. Despite making no real contributions to our research and instrument programs, Nikolay acted like he was our team leader who had a right to our radar systems and resources. But still, we viewed him as a good scientist, and we did our best to work with him.

Terry and I agreed to stop working with Nikolay in 2018. I was near completing my PhD and was preparing to focus my postdoctoral research on climate change. Through our collaborations, Nikolay knew that Terry and I wanted to put an ionosonde at the southern tip of South America. In that region, we could observe weather and climate effects from the Southern Ocean and weather effects from Patagonia, we could partner with research stations on the Antarctica Peninsula, and we could study ionospheric and geomagnetic physics that are unique to the region.

The Air Force Office of Strategic Research has a program that allows university researchers to propose projects that would result in hardware procurement. Nikolay wanted to leverage that opportunity to build a portable ionosonde, and to that end, he convinced us to write a proposal to build an ionosonde in Tierra Del Fuego, Argentina.

Once the project was funded, Nikolay started changing the project plan so he could redirect funds to another project. Eventually he undermined the effort so badly it was no longer possible to complete it. At that point, he didn't need Terry and me any longer, and he fired us from the project. He repurposed the money to build a portable ionosonde.

Abandoning the Tierra Del Fuego project caused harm to a number of our colleagues and stakeholders. Throughout the process, fellow scientists and field site technicians were helping us plan and were relying on the project for various reasons. In hindsight, we believe that Nikolay never intended to deploy the instrument and intentionally wasted everyone's effort so it would give him cover to steal the funds for another project.

Terry complained to the university that Nikolay was committing fraud by repurposing Air Force grant funds. The university ignored the complaint and threatened Terry with disciplinary action for filing a false report. They didn't threaten him because the report was false but because it undermined the university business model of allowing professors to act corruptly.

To add insult to injury, the portable ionosonde Nick wanted to build was based on the intellectual property Terry and I developed and were in the process of trying to patent. We showed him our prototype in 2017 when we deployed it to research a total solar eclipse. Nikolay stole the idea from us, he defrauded the government to build the instrument, and he ruined a ten-year research plan that I had developed to study climate change. Terry and I considered his actions intolerable, and we agreed we would never work with him again.

Despite having disowned him, we did not cut off his data access, which he continued to use for his research. We benefited from access to the data products that were derived from his software, and we didn't have any reason to stop that. We allowed the status quo to continue, but we made sure that we didn't have any obligations to Nikolay whatsoever.

A few months after firing us from the Argentina project, Nikolay had the gall to ask us to work on a new project with the Defense Advanced Research Projects Agency (DARPA), using "his" new instrument. Within one day, both Terry and I had emailed him to formally decline the offer.

Despite the fact that Terry and I refused to support the new proposal, Nikolay submitted it, and DARPA funded it. What we were not told was that he promised DARPA he would conduct research that required use of our instruments at Wallops, Virginia, and Cayey, Puerto Rico. The promise he made was to perform experiments with instruments he had no right to use. That was a major breach of ethics.

In early 2021 Terry and I traveled to Wallops and Cayey to support an Air Force rocket launch campaign. We reprogrammed the two radar systems into an experimental mode of operation. Almost immediately Nikolay began emailing us to ask what we were doing and if he could work with us. The project was a sensitive military experiment and information about it was protected as Controlled Unclassified Information (CUI). CUI was basically a way for the Air Force to allow essential scientists like Terry and me to work on a project without having security clearance. It was explicitly not allowed for us to discuss the experiment, and during the two weeks of the campaign, we simply ignored Nikolay's requests.

A couple days into the project we were contacted by another colleague who told us that Nikolay was asking around about what we were doing. We believed he knew we were working on a military experiment, and he may have been unlawfully attempting to learn classified information. I notified the Air Force Principal Investigator (PI) who responded by asking me if we needed to send someone to "knock on his door." I told the PI that I didn't think that was necessary, and he ordered me to immediately remove Nikolay's system access.

Soon after, we were contacted by the CIRES Director Abdalati, who was asking us why we suspended Nikolay's system access and why we were not responding to his emails. By then, the experiment had concluded, and I had already restored his system access. We explained to Abdalati that it was the decision of the Air Force, and the issue was now resolved.

A few months later we returned to Wallops to support a NASA rocket launch. When we first arrived, we reprogrammed the radar into an experimental operational mode. A couple days later, the Navy created a large explosion off of the Florida coast, and we hoped to research its effects on the atmosphere. The next morning, our computer system logs showed that Nikolay had attempted to install system software on the radar's data-analysis computer, and he was trying to download the experimental data we had collected. Terry and I concluded that Nikolay posed an imminent security risk, and we agreed to suspend his access indefinitely. After all, this was a dangerous radar system on a NASA rocket launch facility, and based on how he was behaving, we had legitimate suspicions that he could be an active Russian agent.

Nikolay sent a series of emails to Terry demanding that Terry restore his system access. Terry replied with one word—"no." Within days, Nikolay convinced CU Dean of Engineering Massimo Ruzzene to ask CIRES Director Waleed Abdalati to seize control of our ionosonde program and give Nikolay administrator-level access to the radar systems. This was insane; Nikolay was never allowed control of the radar systems, and he only ever had a user account, not administrator access, on the data-analysis computers.

Abdalati's unexpected and unilateral decision to seize our program was a direct affront to any reasonable expectation of professional courtesy and internationally recognized academic standards. No respectable academic institution would ever allow such an act.

Abdalati's decision to seize our program was hostile, inappropriate, and made no sense. His actions were basically a response to our efforts to prevent an act of espionage. In hindsight, Terry and I were obligated to report that to the government as a possible act of espionage, but we weren't thinking clearly, and we wanted to do everything we could to resolve the conflict.

After years of doing field work for NOAA, this was the first work trip that I had brought my family to. We had a rental house on the beautiful Virginia Eastern Shore, and my little boy had never seen the ocean. This was a work trip, but our team agreed that I would be allowed to have the Fourth of July weekend to spend with my family. On that Friday, Waleed Abdalati effectively canceled my holiday, and he began a sadistic campaign to destroy our careers and our reputations.

Abdalati followed up with an email that claimed Terry and I had agreed to conduct scientific research with Nikolay. He was specifically referring to the DARPA project that Terry and I both told Nikolay we would not support. Instead of asking us to work toward a solution, Abdalati asserted that the university owned the two radar systems, and as director of the CI, he could do whatever he wanted with them. He said he was taking over control of the systems and ordered Terry to assist with Nikolay's research effort.

One major problem with Abdalati's claim was that CU didn't have any records to show that the university owned either of the radar systems. This was clear because Terry deployed the systems when he worked at the Air Force Research Laboratory, and CU had nothing to do with the project.

Nikolay also knew the radar systems were not owned by CU, but he lied and told Abdalati that they were. Nikolay was on the team that originally developed the systems as part of a Small Business Innovative Research (SBIR) effort, which meant the systems were either privately owned or owned by the federal government.

Abdalati's attempt to seize our program was an outrageous action for multiple reasons. America's prized academic institutions are founded on the principles of academic independence and tenure. Simply put, universities facilitate independent scientific research, and they are not allowed to interfere by telling professors what they can or can't research, whom they have to work with, how to conduct their research, or what specific research to perform. Tenure is a safeguard to make sure that professors are not fired for doing unpopular research. These are basic principles that ensure scientific integrity, and they are fundamental to the idea that we can trust the science.

Seizing our program to force us to support research we never agreed to is very different than a university exercising its authority to enforce obligations that were made to fellow professors or students.

Another mind-blower was the level of extreme indifference to security exhibited by Abdalati, formerly the chief scientist of NASA. He ordered us to give complete control of a dangerous, high-power radar system, on a NASA flight facility, to a Russian expatriate accused of suspicious activity. That cannot be stated enough: The former chief scientist of NASA ordered us, under threat of termination, to give a potential Russian spy full access to a dangerous and high-powered radar system that is located on a NASA rocket launch facility.

The next day was a mess. Terry, who has heart problems, thought that he just had his life's work stolen from him. He became dangerously sick and locked himself in his room for the entire day. I went to the base to work on antennas, and I was so distracted I fell off of a ladder. On that holiday weekend there was nobody around to find me if I had actually gotten hurt. Field work is inherently dangerous, and Abdalati's heavy-handed use of his power had stressed us out, putting both of us in physical danger.

Terry allowed me to take charge of dealing with the issue. I told Director Abdalati that I was trying to help and expected that I would not be punished for it. I let him know that the instruments did not belong to CU, that we did not agree to work on Dr. Zabotin's research project, that Zabotin was a security risk, and that the timing and nature of the demand created an unsafe and hostile workplace. I spent hours carefully wording the email to make sure I respected his authority while setting necessary boundaries.

Waleed Abdalati, probably the most powerful person in NOAA, responded hastily within minutes. He did not take time to think about what I wrote, and he didn't feel it was necessary to use care in responding to one of his senior scientists who just identified a possible national security risk, safety violations, and possible workplace violence. Abdalati was one of the most powerful academics in the world, responding to a serious situation with what seemed to be the care of a high school sophomore. At that point I had no doubt that I was dealing with a narcissistic sociopath who was enraged that his authority was being questioned by an employee he had been trying unsuccessfully to fire for several years.

In his response, Abdalati made the wild claim that ordering a professor to perform research, after the professor had declined to do the work, was not a violation of academic independence. Without any evidence, he falsely claimed that Terry had signed an agreement to do the research. He simultaneously argued that claiming to own the radar systems was not a legal claim and that ownership meant he could do whatever he wanted with the systems.

Claiming he could do whatever he wanted with the radar systems was an absurd legal statement considering that ownership and right of use are not synonymous. Any landlord who has ever tried to evict a tenant knows that you can't throw someone out just because you own the house. This is especially true in an academic environment, where we had funded projects the instruments were already obligated to support.

Abdalati also falsely accused me of making unsubstantiated allegations against a fellow colleague. Everything I wrote about Nikolay was true and accurate and supported by an extensive archive of emails. Just as the university threatened Terry for filing an ethics complaint against Nikolay, Abdalati was threatening me for exposing the truth that put the corrupt university business model at risk.

I had no choice but to cancel my family's holiday weekend and go through years of emails to give Abdalati the information he demanded. I put together a detailed seven-page timeline of events that showed how Nikolay had acted unethically, committed fraud, stolen our research and our intellectual property, and engaged in suspicious activities (see appendix A). The detailed timeline showed how and why we disowned him as a colleague and how and when we explicitly declined to participate in the DARPA project. That document contained an overwhelming preponderance of evidence to show that Terry and I acted professionally and that Abdalati had been lied to and duped by Nikolay.

That document should have prompted an investigation into Nikolay's behavior and led to a severe reprimand or even dismissal from the university. Additionally, it should have forced Abdalati to report possible espionage and fraud to the US government. Instead of taking the action that he was required to, Abdalati decided to destroy my career.

A Satanic Attack and the Federal Response

I believe that good and evil are real, and I view them as tangible things. It follows that there is a struggle between good and evil and people who are immoral or amoral are easy targets to be recruited by the devil. I don't expect anyone else to share my beliefs, but I submit that doesn't matter because we are all capable of seeing right and wrong for what they are.

Was I to believe that we were under a demonic attack or that our senior leadership intentionally acted maliciously and vindictively toward subordinates who were vulnerable to them?

Whatever the case, Abdalati went nuts. Despite giving up my hard-earned family holiday to put together a seven-page timeline that included every bit of information I thought he was asking for, he accused me of not answering his questions, he accused me of dictating who he could discuss the matter with, he claimed that my safety concerns amounted to dictating when we could discuss the matter, and he claimed that his taking control of our real-time weather instruments amounted to a "pretty liberal interpretation" of his demands.

I replied and told him I was only trying to help.

He demanded that we hold a meeting as soon as we returned from our trip. Since his actions appeared unhinged and psychotic and he was disregarding the facts, basic ethics, our legal rights, and any sort of due process, I requested that my wife be allowed to attend the meeting as my advocate. University policy explicitly stated that an advocate would be allowed in this sort of situation, but Abdalati told me she could not attend because she was a federal employee. That made no sense at all.

He disputed my claim of an unsafe workplace, and he accused me of being "demanding, disrespectful, sometimes poorly informed." This was sort of bizarre, considering that I gave him a detailed timeline, completely supported by email records, showing that everything I had said was factually correct and that he was stating things that simply were not true. I knew this was a very powerful and very crazy person I was dealing with. Terry and I were certain he was trying to fire us and hand our program over to Nikolay, and I was terrified.

I followed up by explaining why I felt his actions created a hostile and physically dangerous workplace. I was blunt, expecting that workplace laws and university policy would force him to back down. I cited over a dozen things he did that were interpreted as hostile. I let him know that I worked hard and took great care to respond to him and that his responses seemed to be made with haste. I pointed out that several high-level university administrators, including the dean of engineering, the university research integrity officer, and himself (the CIRES director), had conspired to take control over our research program, without respecting our due-process rights.

When the meeting was held, I answered Abdalati's questions truthfully and respectfully. Despite that, he proceeded to write a letter in which he threatened to fire me for insubordination (see appendix B). In his infinite wisdom, he had accused me of being a repeated problem, citing a reprimand he gave me in 2018 that accused me of "inappropriate and unprofessional behavior." The reprimand he was referring to was the one he emailed me after the whistleblower incident with Juan Rodriquez.

Abdalati's threatening letter was insane. In addition to explicitly citing a whistleblower complaint as one of the reasons he was disciplining me, his rant was dishonest and untruthful.

His comments were revealing, and now I finally thought I understood where that crazy chain of command came from. Abdalati and Kihn concocted it in 2018 as a lever to have their subordinates, Juan and Christine, boss me around. There is a chain of command in the federal government, but there is no such thing for university professors. To protect scientific and academic integrity, the university has levels of administration, not an infinite chain of supervisors that have the authority to scrutinize every decision a scientist makes. It was all quite bizarre but helps show some of the inherent problems with having university faculty work for the federal government.

Abdalati falsely claimed that I failed to recognize his authority as director. To this day, I have no idea what he meant by that. I can only laugh thinking of the cartoon character Eric Cartman from the show *South Park*, yelling, "Respect my authoritar!"

It also bothered me that despite almost two decades of dedicated service with no disciplinary action whatsoever, he claimed that since the 2018 whistleblower complaint I had not behaved according to his expectations. Frankly, I hadn't known he had a list made up for me. I had assumed that obtaining my own funding, meeting my professional obligations, and conducting groundbreaking research was enough to meet any and all of the university's expectations.

Another thing that was strange is that Abdalati referenced a complaint made by Nikolay Zabotin, who was working in the CU engineering department. Despite multiple requests, we still haven't been told what that complaint was. Apparently, CU has some top-secret classification of ethics complaints that are made by one professor on another.

Abdalati claimed that my role as moderator was in lieu of Terry's authority, but he didn't seem to care about the fact that Terry assigned me the task because the situation caused him a serious medical condition. He also accused me of using the unsafe workplace as a lever to delay the meeting.

In the reprimand, Abdalati also accused me of an "unwillingness to respond to very straightforward and simple questions." That was dishonest because the seven-page document I provided was a detailed timeline, fully supported with an extensive email record. I could not understand how my tediously working through years of records to produce a detailed timeline was in any way an improper "unwillingness to respond."

Abdalati seemed particularly annoyed that I wouldn't tell him who owned the radar systems. I tried to explain that he was making a legal claim to property that someone else owned. I had no business providing him with information he could use to assert that claim. This was especially important because the rightful owner of the radar systems had no way to defend against a lawsuit from the behemoth CU legal team. I was not going to help him assert a false claim to that property, and the university needed to do its own research if they wanted to pursue ownership of the radar systems.

In truth, the radar was developed under a Small Business Innovative Research project (SBIR), which is a government program that funds small companies to bring novel products to the market. For these projects, the government retains the rights to the technology, and the small businesses develop a product they can sell commercially. The VIPIR radar systems were the property of a very small company that was run by an elderly scientist who had one part-time employee. If the university were to press a property claim, even if it was baseless, they could use their massive legal resources to effectively steal the hardware. I was not going to help them do that.

The reprimand letter was truly insane and entirely dishonest. I appealed to the vice chancellor for research, Terri Fiez, but as I have come to expect, she buried the complaint and sided with the unhinged CIRES director.

Soon after, I wrote to Terry and told him that because of Abdalati's bullying and harassment, I couldn't perform in the role of moderator any longer. Terry took over and attempted to explain to Abdalati why we felt Nikolay was a security risk. He cited Nikolay's snooping around federal facilities, his attempts to obtain sensitive information, and the ongoing erosion of trust. Terry even went into some of the gory details of computer security at federal facilities.

The following week, Abdalati issued a fresh set of orders that boiled down to a demand for Terry to break federal law and put national security at risk, under the threat of termination. He again ordered Terry to hand over full operational control of a powerful radar system that could harm sensitive electronics on the NASA base to a possible Russian agent who is a world-leading expert in signal intelligence. To highlight how dangerous the order was, if followed, it could have handed the Kremlin control of technology that could be used to destroy spy satellites while they are being launched.

Terry tried to deescalate the situation, but Abdalati was enraged and continued to press harder and harder. He told Terry that he was issuing orders and not negotiating. But neither was Terry, who had no authority to negotiate national security and would not, under any circumstances, be able to comply with the illegal order.

With Abdalati continuing to escalate his threats, Terry and I decided we had to report the incident to NOAA federal supervision. Truthfully, it should have been reported much sooner, but we weren't sure what was the right thing to do. I penned the email, and Terry sent it to our federal supervisor, Rob Redmon.

> Rob,
>
> There is an ongoing security incident where a non-Government person is using a position of authority to forcibly obtain root level computer access for another non-Government person who has previously been identified as a security risk. The computer systems are on US Government property at multiple agencies, and may be connected to US Government networks. There are red flags that raise concerns of possible espionage. Please give me a call on my cell phone ***-***-****.
>
> Terry

Terry then replied to Abdalati and stated that he could not give system access to unauthorized persons.

Terry's email to Rob clearly implied that an act of espionage might be occurring across multiple federal agencies. This should have triggered a protocol that would have initiated an investigation according to a well-defined procedure. For the interest of national security, that is what has to happen, even if the report was made by a low-level employee. In this case, the report was made by two respected senior scientists with nearly a half century of combined service in the federal government.

We believed that Waleed Abdalati used his position of authority to conduct an act of espionage. We were already convinced he was an agent of some special interest and he was doing everything in his power to force us to give a suspected Russian agent unauthorized access to secure systems on government facilities. He did this with no regard for how dangerous Nikolay could have been to US national security and with wanton disregard for our repeated warnings that he may have been a foreign agent. At minimum, Abdalati knew that Nikolay was a Russian expatriate and a world-leading expert in radar technology and the corresponding field of signals intelligence.

Rob forwarded the report to Lt. Commander Bryan Begun, the NCEI executive officer. Commander Begun replied two hours later, acknowledged the seriousness of the accusations, and claimed he consulted with NCEI IT chief Scott Hausman. This was the very same Scott Hausman who ordered the historical climate data to be haphazardly boxed up and shipped to North Carolina. Scott was now our information technology chief, and he was going to sweep these accusations of espionage under the rug. Scott conducted a "thorough" investigation during those two hours and determined there was no threat.

According to Scott, in those two hours, NOAA leadership concluded that the computer systems in question were not within the boundaries of federal security controls according to the Federal Information Security Modernization Act (FISMA). This was a bit of a surprise because we didn't tell them which computer systems we were referring to. We only told them that the systems might be connected to US government networks, which would imply they were within the FISMA boundaries. Without any proper investigation, it was concluded that there was no valid security risk.

In Commander Begun's response to Terry, he copied Waleed Abdalati, who was not a federal employee and was the person who was committing the alleged act of espionage. Lt. Commander Begun unquestionably broke federal law by communicating with outside parties without following the established procedure. In another violation of protocol, he included several federal employees, who may have been accomplices.

What if Terry and I were right and this was an act of espionage? NOAA federal leadership would have given a spy the information and resources he would need to avoid arrest and prosecution. They also revealed the identities of the employees who made the report.

What the lieutenant commander should have done is follow established government procedures for handling the report. Instead, he aided and abetted an act of espionage and put our national security at risk. Why act with such haste and conclude the matter in two short hours? The only conceivable answer is that he did it because he was given no choice. He did it because NOAA leadership reports to and follows the orders of the civilian Waleed Abdalati.

Clearly, Lt. Commander Begun knew what he did was wrong, and he was unhappy that he was put on the spot. The obvious truth is that he was forced to act in the heinous manner that he did. In his show of displeasure, he made it clear to Terry that in the future, he should direct NOAA issues to our division chief, not himself. We found that odd since he was the executive officer who is responsible for handling security incidents such as this.

We later found out that Eric Kihn was the first federal employee who told Abdalati about the complaint. Eric kept himself off of the group emails and contacted Abdalati privately so we wouldn't have evidence that he was behind the corrupted federal response. We only found out because during a legal proceeding several months later, Abdalati provided Eric's email during evidence disclosure.

The following week, Scott Hausman also forwarded Abdalati the exact text of the security complaint. So, a senior federal officer and two senior federal supervisors were notified that their organization was being used for a possible act of espionage, failed to follow established investigative procedures, dismissed the complaint, and tipped off the alleged spy giving him an opportunity to cover his tracks.

When the DOC inspector general's investigators interviewed me later, they concluded that the report of possible espionage was a whistleblower action. They identified Terry as a second whistleblower who had been retaliated against, and expanded the scope of their investigation. But when they issued their report, they omitted all mention of this incident, sweeping another serious matter under the rug.

Burning Bridges

At this point the absurdity had become monumental, and Nikolay's research project was entirely irrelevant to Abdalati. We had established that Nikolay solicited the DARPA project after Terry and I, who are essential personnel, told him we would not participate. When a professor commits this sort of fraud, there is a simple answer: The university tells the sponsor, DARPA, that we're sorry but the professor can't do what he promised, and they return the money. This is not that uncommon, and it's really not a big deal. The attacks on us had nothing to do with the research project.

Abdalati's response was to attack our professional reputations by making an epic scene with our partners. A scientist's reputation is the entire foundation of their career, and he did everything he could to ruin ours.

The first thing he did was write an email to seven federal supervisors at NOAA and told them about a reprimand he issued to Terry. The university reprimand had nothing to do with federal business and these federal employees had no business knowing about it. The only purpose of Abdalati's email was to disparage Terry and harm his standing and reputation within NOAA. This was an overtly malicious act.

Next, Abdalati tried to bypass us by going directly to our partners at USGS and NASA to see if they would give him system access. So, after the forced attempt at access was identified as a possible act of espionage and dismissed by NOAA, he pressed on and attempted to get other federal agencies to give him access. This was another overt, intentional security violation that validated our concern that it could be an act of espionage.

A computer technician at USGS told him, "Any person operating a computer system on DOI/USGS networks must be properly credentialed by having a DOI smart card used for access to systems."[ix] At this point, it is on the record that the system was, in fact, within the boundaries of FISMA. The government also made it clear that the system could only be accessed by someone who had been issued a USGS smart identification card. Terry was an authorized user but neither Waleed Abdalati nor Nikolay Zabotin were. This didn't stop Abdalati from trying to gain access.

Next, he went to the program chief for the USGS Magnetometer program. When she wrote to Terry and me to ask what it was about, we gave her an overview and let her know that Abdalati was on the warpath. I advised her to put the kibosh on his inquiry or he would pester her relentlessly. She took my advice and promptly put an end to Abdalati's inquiry.

In the meantime, Abdalati and his subordinate Ted DeMaria were harassing senior leadership at NASA. They were demanding that NASA

---

ix DOI/USGS refers to US Department of Interior, United States Geological Survey.

figure out who owns the radar systems and grant them access. Abdalati did his best to leverage his status as former NASA chief scientist to no avail—the good people at NASA were having none of it. After more than two hundred emails, they politely insisted that they could neither tell him who owned the system nor give him access to the system.

The people at NASA did the right thing, even at great risk. Abdalati has shown all the markings of a vengeful DC swamp creature and was probably in a position to get back at them through his status within President Biden's administration. Wallops was one of the last bastions of real rocket science, a NASA program that Abdalati himself worked to dismantle. I'm sure these facts weren't lost on the good people at NASA.

Frustrated, Abdalati finally issued Terry a formal reprimand and threatened to fire him (see appendix C). Again, he cited a whistleblower complaint as a reason for the reprimand. In the letter he said that Terry filed a false complaint to NOAA so he could avoid following his orders, which was an outrageous and false accusation. He also implicated Eric by stating that Eric told him he thought Terry's complaint was fabricated to avoid following his orders.

The rest of the letter was an insane egocentric rant using similar language and accusations that he put in the letter he sent me. One of the real gems was an accusation that Terry's refusal to break federal law "is not appropriate, is not acceptable, and unfortunately, continues to damage CIRES relationship with university peers and constituents." And there you have it: Abdalati directly admitting that the university's relationship with NOAA is based on illegal activity.

The next attack Abdalati committed against us was to direct university employees to hold up approval of our primary contract with the US Space Force. He wasn't concerned that the program was important for real-time weather model operations, integral to national security and the safety of our men and women in uniform. He directed Ted DeMaria do his dirty work and interfere with the contract negotiations. Months later, the university's primary contract officer let us know that he didn't like what was going on and apologized to us.

While all of this was going on, the stress was taking its toll on Terry's health, and by the end of September Terry ended up in the emergency room. As we should have expected by now, Abdalati thought Terry was faking it to avoid following his orders. In the meantime, I had to take back the lead role in dealing with the mess. The stress was also deteriorating my health, and that was exacerbated by a lifelong neurological disorder that I suffer from. Since I was a young boy, I've had Tourette syndrome, which causes a movement disorder. The motor ticks that I get are stress-induced, and they create a negative feedback loop that increases my stress, which in turn increases my ticks. Eventually I get muscle fatigue and even physical injuries from the repetitive movements. The situation had made the condition the most severe I had ever experienced, eventually forcing me to seek treatment after decades of being able to manage it myself.

This time I tried to diffuse the situation by allowing Nikolay supervised access to the system. That was a mistake because it just led to more accusations that I was trying to interfere with his work, and more threats from Abdalati. Nikolay showed his appreciation by labeling my effort a "sophisticated mockery."

Behind the scenes, Abdalati was conspiring to undermine our research program with the dean of the engineering school, Massimo Ruzzene, and University Research Integrity Officer John Reuter. Terry and I had some suspicion that these guys were using their positions of power to plot against us, and that suspicion was confirmed when Ruzzene accidentally sent me an email that was intended for Abdalati. Ruzzene wanted Abdalati to take care of me, and Abdalati assured him that I was his guy and he would get it taken care of. Eventually, Ruzzene would be promoted to vice chancellor for research for the university. One of Ruzzene's first acts as vice chancellor would be to fire me.

The university would do nothing to help. The vice chancellor dismissed our complaint; Human Resources completely ignored our pleas for help to deal with a hostile and unsafe workplace; the university ombudsman told us he was sympathetic but we're basically on our own. The university was corrupted and was being paid hundreds of millions of dollars per year to let Abdalati run roughshod over the system.

The only option I had to stop the madness was to file a lawsuit against Abdalati.[25] After being served the summons, he immediately backed off from his overt attacks and redirected his efforts to attacking us covertly.

Removed from the Building

The next scheme was to get us tossed out of NOAA. Abdalati had the university hold up our contract renewal with Space Force, and the project funds had temporarily lapsed. This normally wouldn't be a big deal because we had other projects that we were pulling salary from. Our access to NOAA was to support the NOAA mission, and our positions at NOAA were not dependent on having current Air Force and Space Force funding. Whether or not a specific project was funded was irrelevant, but apparently Abdalati dictates what the federal rules are, and he made it relevant.

Abdalati waited for the project funds to expire and immediately contacted our NOAA federal supervisors. He told them our project was not funded and instructed them to remove our access to the NOAA mission computer systems. Eric Kihn gleefully directed his federal subordinates to follow that order, setting off another circus of retaliation.

The university informing NOAA that a project was no longer funded was unprecedented. Not only has this never happened before, but also there is no reason for anyone to care other than the guys who would not be getting a paycheck. There is no system access requirement that a specific project has to be funded for approved affiliates, who had been thoroughly vetted by the FBI and issued government identification cards. Abdalati and Kihn just couldn't let it go, and because we were on the right side of every issue, they had to invent novel ways to attack us.

Eric ordered his deputy Stephanie Herring and our division chief, Laurel Rachmeler, to use Abdalati's notice as a lever to remove our access to the NOAA computer systems. Because there was no such rule, they just made one up. Stephanie and Laurel called a meeting with Terry and myself, and explained that they were going to remove our computer systems access until our project funding was restored.

There are many requirements for us to have had access to the NOAA mission computer systems. We had to keep up on all of our security training, we had to pass regular FBI background checks, we had to use our government-issued smart cards, we had to maintain our affiliation with the CI, and most importantly, we had to be supporting the NOAA mission. The idea that a specific project had to be funded was ridiculous. In reality, we are allowed to have system access even with no pay at all. We could even be a principal investigator or a program leader and not get paid for it. They just made the rule up because Abdalati told Eric to do that, Eric told Stephanie, and Stephanie told Laurel.

During the meeting, I told Stephanie and Laurel that this was a whistleblower attack coming from Eric and Abdalati and there was no reason to have our system access removed. I also pointed out that our program was supporting NOAA's mission and its mandate under the Code of Federal Regulations (CFR). We also pointed out that disabling our access would put the real-time weather data feed at risk and could compromise national security. They seemed to realize that we were right and decided to take the issue back to Eric.

Shortly after, Eric had them tell us to write a letter to justify our continued systems access. This was also a completely made-up requirement, and a silly one at that. Why would a federal supervisor want to make a paper trail that shows them making up rules to go after a couple of whistleblowers?

But they asked for it, and I wrote a letter that restated what I told them in our meeting. The rule was made up, they didn't have to remove our systems access, the effort was a new line of whistleblower retaliation tactics, and the attempted removal of system access was maliciously jeopardizing national security.

Up the chain it went. This time, Eric apparently felt the need to elevate it to his superiors who in turn reported the letter to the DOC inspector general (IG). The DOC IG identified the DoD as a stakeholder and referred the matter to the Defense Department inspector general. DOC referred the matter back to NCEI to investigate themselves; of course, NCEI swept it under the rug and claimed they did nothing wrong. I don't know how the DoD handled the matter, but our system access was not removed.

The extremely hostile workplace continued for many months. Eric stopped Laurel from hiring me for a job that I was the most qualified candidate for, and he even went to other center directors within NOAA to make sure I wouldn't be hired at another NOAA organization. Eventually I wrote Eric a letter that summarized our history together and that he had promised me a federal job but was instead interfering with my career progress (see appendix D). I finished writing the letter on the last work day of the year, and to make sure Eric received the letter before New Year's, I dropped it off at his house, which I had visited dozens of times over the years.

In the letter I told Eric that the IG would soon release their report and we should work together to take proactive action to solve our problems. I told him that he should hire me to a currently open position so we can show the IG that we did what we could to minimize the damage caused by his retaliatory actions.

Eric gave that letter to Abdalati, who gave it to Ruzzene with the orders to fire me. They had completed the work to destroy my career that began a decade earlier.

CHAPTER 8

THE PURPOSE OF GOVERNMENT

To understand the importance of my story, it needs to be put into the broader context of the purpose of government. What I was subjected to demonstrates an extraordinary level of dysfunction and corruption at NOAA. When government agencies are this far off of the rails, they aren't providing the services that we require of them.

A fundamental role of government is to maintain and protect our society and to prepare us for the challenges that we are likely to face. If we are facing an imminent climate change catastrophe, the primary role of government must be to mitigate the likely consequences of that.

To that end, should our government be putting carbon restrictions on the innovative companies that are developing our next generation green technology? Should the government limit basic research funding for our next generation of green technology, in favor of subsidizing the production of wind and solar farms that provide little or no real progress toward a solution? Or should our government be spending money on infrastructure and development that will help us withstand and mitigate the effects of climate change?

Why are more of us not questioning if our government's priorities are in the right place? One reason is that the climate fascists have control of our most influential institutions, and they have used that to control the rhetoric. Effectively, people are left with little or no means to question the government's policies in a country that is supposed to be a democracy.

The American people don't control our government, they have a dwindling influence on government policy, and the global climate change community is corrupted and is managed by bad people. We need to have an honest discussion and evaluation of where we are as a nation and where we need to go. If the climate catastrophe is approaching as quickly as some are claiming, we need to get this sorted out really fast.

The hard reality is that our government has been hijacked to promote the globalist's climate change agenda. One tragic consequence is that the government is being derelict in its most important job—to protect us.

A Secret Plot to Repurpose a Government Institution

Repurposing government institutions to act as an arm for special interests is complicated. Even as an insider, it took me an entire career to realize how this was being done right under my nose.

As my workplace environment deteriorated, eventually becoming outright hostile, I started piecing together what was really going on. Objectively, over a fourteen-year period, I was an outstanding civil servant. I was well liked by my colleagues, with an excellent history of achievement, and I had never been formally disciplined. Being honest with myself, I can't deny that I'd had my bad moments, but I never acted with malice, and I was always loyal to my superiors and to the taxpayer. Unless Eric had some irrational personal hostility against me, it was clear that something else was going on.

While my colleagues and I were faithfully performing our duties, there were others who were more concerned with gaining the favor of NOAA's senior management, in lieu of their obligations to the taxpayer. To that end, some of my federal supervisors were willing to use their subordinates improperly to further unethical goals, and they put a good deal of effort into covering their tracks. Despite those efforts, people in positions of authority tend to be arrogant and greedy, and some of their bad behavior was eventually exposed. But what was exposed is only a small part of what has really been going on, and the American public is still largely unaware of the scope of the problem within the ranks of our government.

The sinister activity that was being enabled by people like Eric Kihn included a scheme to improperly repurpose the NOAA data centers. The underlying goal of that scheme was to steal the power and influence of NOAA so special interests can manipulate and control public opinion and steer the global climate change industry. Schemes like that are typically executed through the ranks of the Senior Executive Service (SES), which is more commonly known as the "deep state." These SES positions are the highest ranks within the "permanent government." The SES comprises career federal employees who usually remain in place from one presidential administration to another, who are administrators who do very little actual work but have power over all government operations. These are the people whose activities occur mostly out of the public eye, and in our government, they provide some of the highest returns to special interests who are able to corrupt them.

The plot to repurpose the NOAA WDCs had extended across multiple data center directors and across several NOAA administrators. This was a long-term effort.

Within the chain of command below the administrator of NOAA and above the center directors are the assistant administrators and their deputies. Dr. Stephen Volz replaced Mary Kicza as NESDIS assistant administrator (AA) in late 2014. The consolidation of the data centers was a priority objective for Kicza, and she handed that priority over to Volz just two years before NOAA consolidated the data centers into one new organization called the National Centers for Environmental Information (NCEI).

The plot was to dismantle the WDCs. The hollowed-out organizations could be used as a shell to allow the CI directors to infiltrate the federal government on behalf of the special interests they answer to. That seems to have been directed through the NESDIS AAs office across multiple presidential administrations. What we don't know is who was giving orders to the NOAA AAs and the CI directors. The AA's are supposed to report to the NOAA administrator, and the CI director is supposed to report to the university board of regents, but in reality, they report to someone or something else. In other words, NCEI and CIRES are both shell organizations that work together to create a gaping portal into NOAA.

Career officials sponsoring deep state corruption is not unique to NOAA. This type of misuse of government seems to be a common means of corruption across federal agencies. For example, it was recently revealed that multiple federal government intelligence officials conspired to undermine President Trump's authority, even though the president is technically in charge of their agencies.[26] That was an extra-constitutional and corrupt act that was done for the benefit of special interests, at the expense of the American people and the integrity of our elections. Consolidating and repurposing the data centers was a long-term deep state objective, and we don't know who is pulling the strings from behind the curtain.

Ultimately, Dr. Voltz's prior background as associate director for flight programs at NASA and Waleed Abdalati's role as the former NASA chief scientist, suggest deep-rooted ties between them. Both were in senior leadership roles at NASA and were moved into their current positions during the Obama administration. Together, they were in part responsible for replacing NASA's space flight program with a climate-change-oriented mission before they moved on to hollowing out the NOAA WDCs to benefit special interests. They accomplished this without anyone seeming to take notice. The question I would like to have answered is, given their common backgrounds, do they both answer to the same master?

The World Data Centers

By 2009 the United Nations' International Council for Science (ICSU) abolished the World Data Centers and created the World Data System (WDS).[27] This was the culmination of an effort that had been going on for many years and had a seemingly legitimate purpose of addressing challenges created by modern technology. When the WDCs were established in the 1950s, we didn't use computers to collect, analyze, and store environmental data; we did not have giant storage arrays like the multipetabyte systems now commonplace; and we did not exchange information over the World Wide Web. The challenges of the 1950s had been replaced by the challenges of the twenty-first century, and that was used as a way to justify changes in how climate data was managed.

What is concerning to me is that they abandoned the mission of data stewardship and replaced it with a mission of "certifying scientific data services." It seems to me that "certifying" data services has the purpose of giving data legitimacy, but I expect the legitimacy of my scientific data to stand on its own merits.

One thing ICSU and WDS failed to address is there was a need to deal with the remnants of the WDCs. There was still a need to preserve historical climate data, to maintain diplomatic relationships that are central to climate data stewardship, and to keep up with best practices of climate data stewardship and science. Years later I asked Dave Clark, the NOAA lead on the effort to create the WDS, why they abolished the WDCs without apparent regard for these important issues. Dave said he believed it was an honest mistake.

NOAA leadership recognized that abolition of the WDCs could be leveraged to put an end to unfunded mandates that were viewed as a nuisance in a world where climate science had already been "settled." NOAA could abolish the mandates and leave in place established, highly respected organizations like the National Geophysical Data Center (NGDC), the National Climatic Data Center (NCDC), and the National Oceanographic Data Center (NODC) so that they could be leveraged in any way the deep state wanted.

Having known for years that ICSU was moving toward abolishing the WDCs, NOAA leadership had been secretly developing their scheme to consolidate the data centers into what eventually became NCEI. They only had to find a way to make the transition quietly, so the taxpayers and the global science community wouldn't take notice.

There was another problem. In 1979 the United States Congress adopted the Code of Federal Regulations (CFR) 15, chapter 950, establishing, by law, the Environmental Data and Information Service (EDIS) as "a major program element of the National Oceanic and Atmospheric Administration."[28] Under the EDIS, the World Data Centers were codified into the laws of the United States as the first federal agency to manage environmental data and information. The CFR explicitly established the National Geophysical and Solar-Terrestrial Data Center (NGSDC), mandated to "acquire, possess, archive, analyze and disseminate geophysical data, as well as ionospheric, solar and other space environment data."

The problem that NOAA leadership had was that our data center was created by an act of Congress, not by ICSU and the UN. When the UN abolished the international manifestations of the WDCs, they did not abolish the NOAA World Data Centers. NOAA leadership had to find a discreet way to repurpose the data centers in the face of this inconvenient law.

This was a problem that was not limited to NOAA. NASA is mandated by CFR 14 part 1201[29] and exists for the purpose of carrying out the nonmilitary space activities of the United States. NASA's mission, as we all know and understand, is space exploration. But today, NASA is largely focused on climate change, which has little or nothing to do with space exploration. This is clearly a problem now, at the dawn of a new era of space exploration led by private companies like Space-X and Blue Horizon. NASA has been knee-capped in this new space race to colonize the Moon and send the first humans to Mars.

Political operatives, including Stephen Volz and Waleed Abdalati, had successfully infiltrated the leadership ranks of NASA, hollowing out the space exploration program, purging NASA of rocket and space scientists in much the same way NOAA purged itself of climate scientists. They effectively helped repurpose the organization to be heavily focused on climate change. Thanks to deep state actors like them, NASA can't adequately perform its principal, legally mandated function. This was no accident.

NOAA leadership in NESDIS, under the direction of Mary Kicza and Stephen Volz, was determined to do the same thing to the WDCs. The objective was to gut highly respected and established organizations so they could be leveraged by special interests to wield the power and influence of the American people, within the climate change industry.

Recruiting a Fool

NOAA leadership couldn't have accomplished their mission without the help of unwitting government scientists. We did the work that provided the cover they needed.

My role in this story dates back at least to when I was assigned my first WDC-A responsibilities in 2007. As soon as ICSU abolished the UN manifestations of the WDC, NOAA leadership began implementing their plan to repurpose the NOAA's WDCs. They knew they would get push-back from the senior federal data managers, so they pressured them into retirement and didn't fill their positions under the fabricated disguise of a fiscal crisis and hiring freeze.

I was asked to take a leading role in several WDC-A programs as a CI contract employee until the hiring freeze was lifted. I was promised that by taking the role I was guaranteed my federal position. Before me, no contractor had ever performed the functions of a WDC-A program leader. By law, a contractor cannot perform the WDC functions because they are inherently governmental.

Several people in NOAA management knew that I was mission-oriented and could be easily fooled into trusting them. MGG division chief Sue McLean targeted me to take over her prior role as program leader for WDC-A for Geomagnetism. My federal supervisor, Eric Kihn, was more than willing to earn favor with NOAA leadership by recruiting me for that role and promising me that I would be given a federal position.

I was a perfect fit for this job. I was loyal to my superiors, and I was not afraid to take risks by circumventing established rules and policies to get the mission accomplished. NOAA leadership knew that my responsibilities were legally not allowed, they persuaded me that the job had to get done, and they promised I would be rewarded for taking the initiative.

In hindsight, I can see they were lying to me from the start. Eric, Sue, and high-level administrators knowingly deceived me so I would do their dirty work. It was clear from the beginning that the guy who knows where the metaphorical bodies are buried could never be made whole and given a senior federal position. I have no doubt that I was targeted to be used and discarded.

What still seems to have been a pretty dumb move is betraying the guy who had proven to be reliable and loyal. My best guess is the orders came down from Volz or whomever he was taking orders from. The orders must have been issued by someone who didn't know me personally, could act without remorse, and could make that decision on an uninformed basis, without realizing how hard I was going to fight back.

The truth is, I wouldn't have guessed they were using me this way, and I never saw it coming. Simply put, I was their fool.

CHAPTER 9

ANATOMY OF A DEEP STATE PLOT

How the deep state executed their plan was secretive and complex. Each of us had our own role to play, but to the extent possible, leadership kept us in the dark. What I experienced was only a small part of the process, but it was enough to help paint a picture of their larger scheme.

I witnessed several key components of the plot. Before the WDCs were consolidated into NCEI, NOAA leadership made sure they had their operatives placed in the WDC director positions. Tom Karl was the director of the National Climatic Data Center, and Eric Kihn would be appointed acting director of the National Geophysical Data Center. Both of these operatives would begin to insert themselves into international climate change politics, and they assembled teams of useful idiots like myself to wind down NOAA's WDC role.

Even though the UN abolished the WDC in 2008, NOAA leadership waited to formally consolidate the data centers. The NESDIS AA finally made the risky move in 2015, near the end of the Obama administration, before a possible Republican administration could come in and disrupt their plan.

When President Trump took office, he appointed Wilbur Ross as commerce secretary. I joked that Secretary Ross was the cartoon character Mr. Magoo because he was an old, seemingly mentally shot political appointee who didn't appear to have much interest in climate change. He also kind of looked like Mr. Magoo.

At the same time, Congress was slow-walking confirmation of the president's appointees, including the undersecretary of commerce for oceans and atmosphere, who serves as the NOAA administrator. With Ross as the new commerce secretary and a leadership void at NOAA, AA Volz was able to continue the transition away from the WDC mission with minimal risk of the administration taking notice and asking questions.

Once the consolidation was complete, NOAA redirected all WDC-A funds to the CI director, put the CI director in charge of all research programs, and redefined how CI employees are allowed to interface with the federal government. Leadership at the newly formed NCEI inserted themselves into international climate change roles and worked to clean up loose ends that could expose their plot.

The Useful Idiots

I took a leading role in the quiet dismantling of the WDC-A in 2009 when I was appointed to be WDC-A program leader. Eric had already known I was hardworking and loyal, and I was identified as the perfect candidate to be one of the useful idiots. Eric directed my division chief to quietly wind down as many WDC activities as he could, and they had me do the dirty work. There was a risk for NOAA leadership because only federal employees could legally serve as WDC-A program leaders, and I was a contractor. Over time I took a more prominent role within the WDC, and that became more and more of a problem for NOAA leadership.

During those years, Dave Fischman, an NOAA corps officer and a close personal friend of mine, gained the favor of MGG division chief Sue McLean. Dave is a dedicated civil servant who not only has always acted in good faith but also has not been afraid to gently challenge leadership when it was necessary to do so. For years, Dave served as a volunteer firefighter, and he is now the chief of his fire department in the mountain community of Sugarloaf, Colorado. Dave is also one of the few NOAA scientists who would volunteer to help when data was at risk. Dave was one of the good guys.

Sue knew that Dave was planning to leave the NOAA corps, and the only way to keep him around was to give him a federal job. The hiring freeze was magically lifted, but for only one position. That position was for the role that I was acting in: program leader for WDC-A for Geomagnetism.

NOAA leadership always has a reason for the actions they take, and they hired Dave over me for two apparent reasons. The first reason was they needed a federal employee to officially hold the title. Since I had been working in the position illegally, I was a liability, and it would be too risky to hire me. Also, this was their way to tell me that they were never going to come through on their promise to give me a federal position.

The other reason they picked Dave was because he was my friend, and they are mean-spirited people who got off on putting a wedge between us. Dave was a hydrogeologist, and he didn't have any of the space physics expertise that was required to do the job. If they lifted the hiring freeze simply to keep Dave, it would have made much more sense to open up similar positions that had been vacated by federal data managers within the MGG division.

Before hiring Dave, leadership only let me act in the role for a year or so because they knew the WDC position was inherently federal and they couldn't allow me to keep acting on behalf of the federal government while being a contractor. By the time the hiring process was completed, I had been acting in the position for two years, I was the most qualified candidate, and I had a stellar reputation within NOAA and among our international partners. I knew that passing me over was a slap in the face, but it didn't occur to me that it was intentionally sadistic.

As soon as Dave was hired, he began working with Sue in the MGG division, and he never actually took over any of my responsibilities. He did support my work the best he could, and I welcomed that, but management continued to have me, the real subject-matter expert, perform all the duties of a WDC-A program leader.

When I protested to Eric, he told me that Sue wanted Dave so she wouldn't lose him, and he said they will move him into another position soon and then reopen the WDC position for me. Like the useful idiot that I was, I took him for his word and continued performing my duties.

Now I understand that by hiring Dave as the official program leader, they had given themselves cover. Having done that, there was a federal employee in the role, and they couldn't be accused of having a contractor perform diplomatic missions for the US government.

Operatives and Directors

By 2012 I was still believing Eric's lies. After all, we were good friends who would spend personal time together, with our families. On Halloween, we would take our daughters trick-or-treating together. I sincerely believed that Eric was looking out for my best interest, but he simply had no power to fulfill his promise to me. When our former director Chris Fox announced his retirement, I begged him to mentor Eric as his replacement. I argued that Eric was the future of NGDC and that having a space weather scientist in the director's chair was long overdue.

I don't know how much influence I had on Chris's decision, but before he left, he mentored Eric, and Eric became his replacement. Honestly, Eric's promotion was probably more a result of his willingness to act unethically to gain the favor of his superiors than any influence I might have had. Eric would end up being a perfect operative in the director's seat; his actions seemed amoral, and he was willing to throw his people under the bus to advance leadership's corrupt transition plan.

A short time after Eric replaced Chris, he started the process of forcing me out of my job. First, he cut half of my salary, claiming that NOAA was no longer willing to fund a single full-time employee to support the WDC-A. That should have been a red flag because for the previous half century, the program always had at least a half-dozen full-time employees, and Congress was still funding every cent of that salary money. I began working for Terry in the ionosonde program, and I was still expected to continue my work running the WDC program. Eric still needed me to provide cover for the WDC, but he also knew that was a risk, and he ordered Bill to slowly push me out of the role. Bill, as division chief, started performing some of the low-level data-clerk functions, insisting that I had to reduce my WDC workload.

I approached Bill and strongly protested what was going on. I reminded him that I agreed to take on the role in good faith, and I objected to being used to destroy an important program, and my own reputation with it. I told him in no uncertain terms that I objected to him putting me at the helm of a sinking ship where the global scientific community would see me as the person responsible. He assured me that was not going to happen.

But it did happen. It was all part of the plan. And then the consolidation happened.

In early 2015, during a four-and-a-half-month trip to Antarctica, Bill sent me an email telling me that NGDC no longer existed and effective immediately NOAA would no longer provide any support for my WDC work. In other words, after ten years of service, Eric ordered Bill to fire me the minute he had an excuse.

Eric revealed his true nature, firing his long-time loyal subject and friend, without notice, in the middle of a long trip away from my family. This was a clear and intentional message from Eric that I had worn out my welcome. Terry immediately stepped up and took me on full time as his protégé.

NCDC Director Tom Karl was another corrupt operative who was used by the AA to facilitate the consolidations. Some of his actions leading up to and following the consolidation were to publish a fudged climate change paper to influence the IPCC before the Paris Agreement, redirect the funds for all WDC-A activities, and remove all historical climate data records from public access and from the stewardship of subject-matter experts at the three NOAA WDCs.

During his time with the National Climatic Data Center, Tom Karl had long been associated with the IPCC and predicting an imminent climate catastrophe, which had been used to push theories like the imminent demise of polar bears.[30] Tom was probably awarded his position as director of NCEI because he was a long-standing insider for the politicization of climate change. But he was focused on politics, not being a good director. By the time Tom Karl retired, the agency was a mess and had no clear mission, and the people of NCEI were demoralized.

In 2017 Karl's work gained national attention when a recently retired NOAA scientist wrote an op-ed claiming that he rushed out a NOAA-sponsored paper to the journal *Science* ahead of an important meeting of the IPCC, to influence the infamous Paris Agreement.[31] It was later revealed that he destroyed evidence that he had improperly manipulated the climate record.[32] This sparked an international controversy that created a political firestorm.

What the investigators missed, and the public couldn't see, was that the effort to influence the IPCC was a piece of a much larger plot that was unfolding at NOAA.

The Aloof Secretary of Commerce

Under the first Trump administration, the secretary of commerce was Wilbur Ross. Secretary Ross was very old and didn't seem to have much spunk. Similar to other commerce secretaries during Republican administrations, Secretary Ross took a very hands-off approach with NOAA and seemed to have focused on the space flight components of NOAA's mission. NOAA had a sprawling space flight program that seemed to be part of a larger pattern of NOAA undermining the weather and climate mission in favor of expensive, high visibility operations. This was comparable to NASA undermining its space flight mission in favor of weather and climate research.

During the entire Trump administration, I recall only two correspondences from the secretary's office: The first was a generic "hello, I'm your new DOC secretary," and I remember the second being related to the breakout of the COVID pandemic and ensuing lockdowns. Ross was pretty much out to lunch and handed the keys of NOAA over to the deep state holdovers. This gave AA Volz the time window he needed to complete repurposing the NOAA WDCs without the administration taking notice.

One of the things my colleagues and I noticed after NCEI consolidated was that communication between leadership and low-level workers had mostly stopped. This helped facilitate a demoralized workplace, and many of my colleagues were complaining that they didn't know what our mission was and were afraid for the organization's future. Government workers tend to get paranoid when they think their organization is going to be mothballed, and without a clear mission, that is what they thought might happen.

Leadership did let us know that the old mission was no longer in play, but they never told us what the new mission was. In hindsight, that was obviously because the new mission was corrupt and improper, and they didn't have a good way to get us all on board.

Consolidation of the Data Centers

When the NOAA WDCs were consolidated into NCEI, NCDC Director Tom Karl was appointed as the new NCEI director and immediately repurposed the organization. I guess he beat out Eric Kihn and the NODC director in their race to be the most effective stooge for the AA. Then again, Tom had the most long-standing history within the ranks of the international climate change fascists, and he was positioned to be the most effective influencer for the newly formed NCEI.

One of the immediate actions they took was to defund all WDC activities. At the time, I had my hands full building a radar system at Jang Bogo Station in Antarctica, so I never did get a handle on what happened during the actual consolidation. All I received was the overnight email from Bill telling me I was out of a job.

It wasn't long before NOAA leadership ordered that all historical climate data records be moved to a single location, in the basement below the federal courthouse, in Asheville. The result of this move was to make all historical climate data unavailable to the scientific community. Even public discovery of the data became all but impossible. By removing data access from experts like me, numerous research projects were harmed, impeding progress on our understanding of climate change.

One of the most seemingly odd transitions was a change in the workplace command structure for CI employees. NOAA leadership told us that we were no longer allowed to report to any federal supervisors. They gave us some bureaucratic nonsense about federal rules, but that was nothing more than a weak cover story. Everyone understood that the entire point of the cooperative institutes was to allow contractors to provide direct support to the federal mission. The reason for that was simple: The federal workforce needed skilled people to perform certain tasks, and it wasn't always practical to have those performed by career federal employees. That need was more important than ever because so many federal scientists retired during the hiring freeze and they weren't being replaced.

When I arrived at NOAA, CI employees were hired through a process in which the federal supervisors would ask the CI to approve and process the hiring. The entire process was under federal direction, and the CI contribution was to rubber-stamp what the federal managers asked for. Once hired, the CI employees reported directly to their federal supervisors and were assigned a CI supervisor to ensure compliance with university rules and regulations. We didn't report to our CI supervisor, and I didn't even know who mine was until I was required to meet him for an annual performance review.

The entire system was set up for the purpose of having contractors work for federal supervisors. Using federal rules as an excuse to end that practice, without warning, was not only harmful to the mission but also silly and dishonest. Clearly there was no new law that forced a change to this practice, and even if there were, it would be a reason to reevaluate or even terminate the CI relationships. Instead, they allowed contract affiliates to operate indiscriminately within the federal workplace. Everything NOAA leadership does is planned and has purpose, and very clearly, this has been part of the plan all along.

One Principal Investigator and Hundreds of PhDs

Another piece of the transition plan was to have the CI director act as principal investigator of all NOAA-funded projects. This was another bizarre move that only makes sense once we realize that he is the unregistered agent who had been assigned to leverage NOAA's power and influence on behalf of special interests.

In scientific research, the principal investigator (PI) is the lead scientist on the project. The PI is expected to understand and direct every aspect of his research program and is usually expected to publish the key findings as first author. A single person could not ever serve in that role for more than a couple research programs, let alone hundreds of them. This was a massive power grab, quite possibly the biggest power grab in the history of science.

The programs of which Abdalati was made PI take in hundreds of millions of dollars of NOAA money every year. That much money wouldn't be unusual if it was for an expensive satellite mission. It is not sensible or legitimate when it is spread out over hundreds of high-impact, million dollar per year research projects. That responsibility was given to Waleed Abdalati, who I have shown makes poor, emotionally driven decisions. The purpose was not to have the CI director manage these programs, but to wield the power and influence of hundreds of world-class NOAA research programs. As PI of these high-impact projects, Abdalati now has access to, and leverage over, the most senior scientists and decision makers in the world.

There was also an issue of the CIRES director having weekly "lunch" meetings with NOAA federal supervisors. At these meetings, NOAA federal supervisors can illegally receive orders from him and take improper government actions at his direction. These informal weekly meetings allow the CI director a way to issue his orders to the federal supervisors in an off-the-record setting that provides cover for illegal behavior.

Understanding the Conspiracy

Simply put, NOAA Assistant Administrator Stephen Volz and NASA Chief Scientist Waleed Abdalati were recruited to hollow out the NOAA World Data Centers and seize the organization's power and influence. Whom they were recruited by and whom they answer to is a special interest that I haven't been able to identify, but we can see the vulnerabilities in the system—and vulnerabilities have a tendency of being exploited.

To accomplish their scheme, they enlisted Tom Karl, Eric Kihn, and other NOAA federal employees to implement a number of changes over a short time period while the secretary of commerce wasn't paying attention. They consolidated the NOAA WDCs into a new organization named NCEI, they defunded all WDC activities, they forced out WDC-A employees, they cut off the rest of the organization from any useful information or guidance on the transition, and they cut off the taxpayer's access to NOAA's extensive archive of historical climate data records.

These deep state actors cut off the remaining CI employees from the federal chain of command, but continued to allow them to work freely within the government walls, without having any government functions to perform. They put the CI directors in full control of all programs and funding, and the federal supervisors were made to follow the orders of the CI director.

The University of Colorado who hosts the CI is effectively bribed to turn a blind eye to all of this. This created a vehicle for the CI director to have the federal government act on his behalf, including the ability to bring people into and out of the federal workplace and to leverage and take credit for all aspects of NOAA-funded scientific research. Furthermore, the federal supervisors were directed to protect the scheme at all costs, including the cover up of potential acts of espionage.

The entire plot carries the primary goal of using NOAA's power and influence to sway the IPCC and other international governing bodies and to steer the direction of the climate change industry. This is an effective way to control our futures, from the cars we drive, to the food we eat, to the homes we live in, to the families we raise.

What is even more scary is that we do not know who Volz and Abdalati work for and answer to. Officially, Abdalati answers to the University of Colorado, and Volz answers to the NOAA administrator. But as I have pointed out, Volz's appointment has outlasted multiple presidential administrations, and the university receives hundreds of millions of dollars a year that is effectively a bribe to hire whom they are told to hire and to turn a blind eye to what goes on at the CI.

CHAPTER 10

EXPLOITING THE SCIENCE

NOAA isn't the only institution that has been hijacked by special interests. How the power and influence of NOAA is wielded by an operative like the CI director is based on an effective, well-developed model. One purpose is to leverage how science works.

Scientists are expected to collaborate and exchange ideas. I study the upper atmosphere in the fields of space physics and aeronomy. If you go to an academic conference in space physics and aeronomy, and talk to a random scientist who studies the mesosphere and lower thermosphere region of the atmosphere, there's a good chance they know who I am. In these academic meetings, we typically exchange scientific ideas and discuss new research opportunities.

Some scientists will attend meetings that cater to decision-makers and private industry professionals. Those meetings are more focused on funding programs and their objectives. This is where we focus on the mission, and the science becomes less academic in nature. To accomplish the mission, we can offer to perform specific lines of research and deliverables. We can agree to redirect our resources and our focus, and those favors can easily be tied to a quid pro quo. This is where policymakers who foot the bills and control the trillion-dollar climate change industry can help the "scientist" understand what results are expected and what results are unacceptable. This is where power and influence can be bartered.

At NOAA, the senior leadership handed the majority of government-sponsored climate science to a single operative, so the government's power and influence can be bartered for the benefit of the special interests that he answers to. The university was bribed with hundreds of millions of dollars per year of taxpayer money, so they turn a blind eye to the clear and obvious betrayal of academic integrity, and federal employees illegally obey the will of a civilian political operative.

The corruption that I have exposed at NOAA is aimed at influencing the global climate change industry. That industry is controlled by political operatives, through the United Nations.

The United Nations

In 2024 we face challenges with climate change that didn't exist fifty years ago. To prevent an imminent global catastrophe, the United Nations is calling for dramatic and immediate action to prevent average global temperatures from rising to 1.5 degrees Celsius above the preindustrial average.

If we could reach any real consensus, it would be to work toward the establishment of sensible and effective policies to mitigate the damage that we are doing to our planet. Earth sciences are vast and complex. Nobody has yet been able to propose a reasonable comprehensive solution, but the framework has been proposed. Many important ideas have even been adopted by the inherently political institutions of the United Nations.

These ideas really begin with an acceptance that our climate is changing and that human influence adversely affects the Earth's ecosystems and the environment we live in. It's important to acknowledge that there are many uncertainties in our understanding of climate change and that the science is far from being settled. We need to be conscious of the importance of continued scientific research, and we should adapt to changes in our understanding of science.

A reasonable approach to tackling climate change can be found using common sense, and the UN's first efforts to deal with climate change were very sensible. As a practical matter, the solutions must be feasible environmentally, socially, and economically. Most obviously, the solutions need to actually help the environment.

The purpose of the UN is having nations work together to create a better world for ourselves and our posterity. To that end, fruitful cooperation isn't possible if the efforts put one nation at a disadvantage with their adversaries. There is also a need to be careful to not cause adverse impacts to society and to consider the legitimate needs of all people. Ultimately, there needs to be equitable sharing of burden and risk. In general, any large-scale international effort should focus on and result in sustained economic growth.

Another common-sense goal has to be a focus on the eradication of poverty across the globe. People who don't have basic needs for survival can't be good stewards of the environment. Only when people can meet their basic needs are they able to concern themselves with the long term and plan for the future. A big part of planning for the future is protecting and sustaining our environment. Eradication of poverty would also keep people from harming the environment to meet immediate survival needs. A family that is freezing to death might need to burn used tires to survive the night, without regard for the pollution that will cause.

Every country has a right to develop its own resources and promote its economy. Each nation needs to be considerate of the rest of us and avoid polluting our environment. Also, nations should be encouraged to enact their own legislation rather than trying to strong-arm everyone into a one-sided policy formulated by global elites. The heavy-handed fascist approach that has been adopted doesn't facilitate independence, cooperation, or consideration.

The reality is that the lion's share of greenhouse gas emissions come from developed countries. Developed countries include India and China, which the UN corruptly considers developing nations so they can produce cheap goods at the expense of a large carbon footprint.

In underdeveloped countries, poverty leads to a disregard for good environmental policies. This has led to an understanding that the most effective climate policy is for developed nations to produce environmentally friendly technologies and for developing nations to focus on elimination of poverty, so the poor aren't forced to pollute to meet basic survival needs. We also need good, continuing scientific research to understand how greenhouse gasses are created, stored, and destroyed.

It also needs to be recognized that some places are more vulnerable to climate change than other places. Most of the world can adapt to a changing climate, but some places can become deserts or be taken over by invasive species or experience severe disruption in food production, like farming and fishing. The needs of the most vulnerable cannot be overlooked.

Initially, the United Nations adopted all of these common-sense policies and solutions, and the job of being good stewards of our environment was done in the spirit of cooperation. The problem is that they did all of this more than fifty years ago.[33] The sad truth is that what began as a good-faith approach to real-world problems has been allowed to be completely corrupted. The UN operatives who corrupted the system have done so methodically and brilliantly, taking advantage of our collective ignorance of science.

Fascist strong-arm tactics like those that have been implemented through the Kyoto Protocol and the Paris Agreement, including the public shaming and penalizing of countries without due process, do not facilitate productive global cooperation. There is no reason to believe that any of the actions recommended by the IPCC can be more effective than an adherence to the ideas the UN embraced in 1972. To the contrary, the current UN policies are causing an increase in poverty and poverty-induced environmental harm. These policies allow countries like China and India to pollute indiscriminately while putting developed Western nations at an economic disadvantage, they hinder technological advancement in favor of subsidizing outdated technologies, and they promote corrupt political deals and division.

The Beginnings of a Global Institution

It is evident that the global elites in charge of the UN climate change institutions are driven by power and influence, not the greater good. Our institutions have become corrupted and abused, and that cannot result in the best outcomes.

Global institutions like the IPCC don't manifest from thin air. Their beginnings are often rooted in goodwill, with a purpose and a well-defined mission. Like most government agencies, they have noble beginnings that are eventually repurposed by those who will use them to wield power and influence.

Very few noble ideas are born in the minds of elitist power brokers. The best efforts in international cooperation in the Earth sciences were started by scientists, not politicians. These ambitious initiatives were taken by scientists who seem to have been passionate about their work and who had a genuine interest in making a better world for all of us. This is what we expect scientists to do when they are not corrupted or blinded by ignorance.

One idea in particular had planted the seed that this story has grown out of. In the early 1950s the world was in a new era of weapons of mass destruction and rapid advancements in rocketry and computer technology. A group of prominent scientists met in the living room of the famous space physicist James Van Allen, put their heads together, and agreed that we needed to have a global conference to deal with how these challenges were affecting our planet. They proposed to leverage and expand upon an existing institution called the International Polar Year (IPY). They decided that at the next IPY meeting beginning in 1957, the International Geophysical Year (IGY) would take place. The IGY would bring together Earth scientists from all over the world to collaborate.

When you and I think of weather and climate, we tend to think about global warming, severe storms, and fossil fuels. In the 1950s our scientists were thinking about the Earth system as a whole, with comparable emphasis on many fields of scientific research, including meteorology, oceanography, seismology, geomagnetism, precision surveys and mapping, gravity, solar activity, and space sciences, including aurora, cosmic rays, and ionospheric physics.[34] The irony is that today, technology allows scientists to be more focused on robust cross-disciplinary research, but politics have driven us to think in the most simplified terms of greenhouse emissions and global warming.

The World Data Centers were established in 1957, right before the IGY. The IGY focused on a comprehensive approach to Earth sciences that resulted in several major accomplishments. It was resolved that "all observational data shall be available to scientists and scientific institutions in all countries." Years into the Cold War between the Soviet Union and the United States, the two nuclear rivals were able to establish diplomatic ties in science, for the purpose of making the world a better place. That tradition managed to persist for almost six decades and ensured the establishment of a robust climate data record.

For the remainder of the twentieth century, the WDCs protected our data, handled and rectified different data formats, and promoted the free and open exchange of data. It was my honor to become a part of this tradition as an officer of international scientific organizations, as the lead for a major WDC program and as a NOAA representative for climate data modernization. Among the greatest honors of my life was being trusted with these great responsibilities by my predecessors.

Stockholm 1972 and the UN Environmental Programme

The WDC system was enormously successful at facilitating international collaboration across all fields of the Earth sciences. We were finally able to piece together the fundamentals of weather and climate, giving the scientific community an opportunity to influence the policies of world governments. It wasn't long before the UN decided to host the world's first political conference on the climate, which was held in Stockholm in 1972.

The principles set forth by the Stockholm conference seem to hit the mark that was needed to put us on a trajectory of resilience against climate change. At this meeting, the UN recognized fundamental human rights that included the need to safeguard resources and protect the Earth's ability to provide for us. They recognized the fundamental truth that it's bad to pollute, and they promoted social and economic development, rational societal planning, and the need to focus on development of science and technology.

The UN recognized that poverty causes harm to the environment. People who are struggling for basic survival aren't concerned with protecting the environment and are willing to harm it to get what they need. They recognized that it is through education that each of us can understand our role in protecting our planet.

It was clear that human activities affect the environment, and the environment affects our well-being. It is the government's duty to protect the well-being of its people.

In these early days, the UN member states recognized that there was risk to implementing change to protect the environment, and they understood that they needed to be prudent and action needed to be taken carefully. They understood that we needed an "enthusiastic but calm state of mind" to achieve the "fundamental goals of peace and of world-wide economic and social development."

A key part of having a measure of prudence was understood to be an acknowledgment that every person and every government play a different role. They correctly recognized that environmental problems in poor places are caused by poverty itself. It was decided that developing nations should direct their attention to the eradication of poverty.

On the other hand, it was recognized that advanced nations could develop the technology needed to facilitate economic growth while protecting the environment. The UN member states recognized that population and economic growth is natural, and we need to work with it. What they claimed to strive for was a better life for ourselves and our posterity while keeping true to the fundamental principles of human needs and hope.

The balance that was identified is the need for all people to share the burden in equity. I've tried to understand what this means, and the only answer I have is dependent on how we might be able to measure equity. It seems to me that equity in burden-sharing would at minimum ensure that no one nation or individual is put at a disadvantage to other nations or individuals as a consequence of their contribution. The concept seems to have been lost in time, with the American people bearing a disproportionate amount of the burden, putting us at an economic and strategic disadvantage with respect to our adversaries.

Following in the tradition established by the IGY, the UN resolved to deal with international matters cooperatively and establish ways for nations to work together toward our common goals. In part, this included a recognition of the need to eliminate weapons of mass destruction, a most basic truth that everyone understands, which our leaders have comprehensively failed to achieve despite living under a threat of mutually assured destruction for most of a century.

The principles set forth in the Stockholm meeting are timeless but have also been largely lost to time. They recognized that there are fundamental human rights, that we need to safeguard our resources, and that we need to protect the Earth's ability to act as a natural engine for renewable resources.

There was a novel principle set forth: polluting is bad. It was understood that the best remedies for pollution included the eradication of poverty and promotion of social and economic development. How could they have possibly understood that rational planning and education with a focus on science and technology could be leveraged to achieve this goal? How is it that more than a half century ago, they understood that we could approach international matters cooperatively?

Or perhaps the more obvious question is, why are these principals no longer central to protecting our environment? How is it that the application of force has become the only way to reach the goals that would be accomplished by cooperation and a mutual respect for our fundamental rights and dignity?

The Stockholm meeting marked the beginning of an international collaboration on climate change management, in a time when world leaders approached our environmental challenges with a measure of sanity.

The United Nations Framework Convention on Climate Change

Between 1985 and 1990, the UN took a number of actions to address environmental degradation, sea level rise, desertification, and the hole in the ozone layer.[35] These were all valid concerns that were leveraged to set the stage for a convention in 1992. During this period, the International Panel on Climate Change (IPCC) was established as an advisory panel for world governments.[36]

During the convention of 1992, the United Nations Framework Convention on Climate Change (UNFCCC) was established. The UNFCCC would quickly evolve into the climate change Gestapo that we know today.

It was a real problem that significant ozone depletion was being caused by manmade ozone depleting chemicals (ODC). This created what is commonly known as the ozone hole. The UN identified the main problem caused by ozone depletion as risk to human health. This led to the

Montreal Protocol in 1987, which banned many of these ozone depleting chemicals.[37]

The result has been a reversal in ozone depletion, and the atmosphere is in the process of recovering naturally. The Montreal Protocol was an example of a successful implementation of technologies to solve an important environmental problem. There were challenges, including economic costs associated with replacement of ODC's that were resisted by stakeholders such as the hydrocarbon industry that produced the chemicals. There were also disagreements between nations, but the international effort was ultimately a result of cooperation, without causing significant harm to the world economies.

Loss of fertile lands and rising sea levels were also an increasing concern, leading to UN resolutions on desertification and sea level rise, and a plan of action to combat desertification.[38] The 1992 convention has the appearance of a sensible application of the 1972 principals to address these issues.

The convention had a stated objective to mitigate environmental damage in a measured way. The intention was to avoid moving too slowly to get ahead of the problems, without moving so quickly we would upend global economics. It was an expressly stated purpose to maintain sustained economic development while working to protect the environment.

The UN also stuck to the principle of fair burden-sharing. The specific needs of countries that bear a disproportionate burden were to be given priority consideration. There was also a commitment to make scientific computations using the "best available scientific knowledge." That seems to have been an acknowledgment that there is a risk created if we rely on incomplete or undeveloped science.

Overall, the stated motivations of the convention were consistent with the principles set forth during the 1972 convention. There was an acknowledgment that the climate is changing and that human activity will adversely affect the Earth's ecosystems and human society.

The approach was still scientific and practical. There was a recognition that there were many uncertainties in our understanding of climate change and that we needed to use the best science available to determine the sources, sinks, and reservoirs for greenhouse gasses. The parties to the convention even had the sense to recognize that poor countries can do their part by reducing poverty while developed nations provide technological solutions.

The convention reaffirmed the sovereignty of nations and the need for international cooperation. They agreed that nations had the right to develop their resources, promote economic development, and provide a better life for their citizens. There didn't seem to be any feeling that we had to sacrifice the now for a theoretical future.

The overall objectives were quite sensible and should still apply today. There was a goal of coming up with solutions that were economical and environmentally effective. It was understood that we had to focus on sustained economic growth and the eradication of poverty. There was a recognized need to have the best and most comprehensive scientific research solve the unknowns. Most of all, there was an understanding that the goal was to protect our planet for both current and future generations.

The Kyoto Protocol and the Paris Agreement

Quite obviously, things have changed within the halls of power. We are now told we need to set aside economic concerns to reach what they say will be effective environmental goals. We are allowing poverty to run rampant with skyrocketing fuel prices, so we can put a small dent in greenhouse gas emissions. We are now told that the science is settled and we need to solve our problems with existing technology. We expend too many of our resources subsidizing existing technologies at the expense of development of newer and better green technologies. Most disturbingly, we are being asked to sacrifice the well-being of our current generation for some unknown future.

When we peel back the layers, it becomes clear that the power brokers in charge of the UNFCCC have been aggressively establishing a monopoly over power and influence in the climate change industry. When we give them the means to solve our problems on pollution, poverty, and climate change, they put their energy into maximizing their power and influence over the entire global economy.

Through the 1992 convention, efforts to use climate change as a lever to manipulate power and influence were kept mostly behind the scenes. That all changed in 1997 with the adoption of the Kyoto Protocol.

The Kyoto Protocol created a legal commitment for countries to reduce their GHGs that was based on agree-upon individual targets instead of fair burden-sharing. Basically, this meant that the more generous nations like the United States took on a disproportionate responsibility. Developing nations did not take on any binding responsibilities under the Kyoto Protocol, but they were given power to force others to abide by the agreement.

Simply put, the central principles of the 1972 Stockholm agreement were thrown out the window and replaced with a fascist rule-making and enforcement system.

The Kyoto Protocol is responsible for the emission trading system that most of us understand as carbon credits. That is a system that allows countries to deceptively offset their emissions through certain qualifying financial contributions. With that, an artificial global economy was established based on what is cumulatively the carbon market. I'm not an economist, but the entire structure of this faux market seems odd to me given that each nation that participates takes on a different debt load that seems to be based on political favors instead of any substantive product value. The participants can trade that debt while being expected to share their technologies as an additional burden. It's not clear to me why this makes more sense than developing new technology in a legitimate market, consistent with the principle of economic feasibility that the UN had maintained for decades.

There is also a strange provision that encourages the private sector to contribute. But in any market system, private interests are only expected to contribute when there is value for them in doing so. Was this an early roll out of a social credit system for the West?

Of course, they had to set up a climate enforcement Gestapo that they called the Compliance Committee. But the committee seems to be mostly run by representatives of the underdeveloped nations that have no obligation for compliance. If we were to have any expectation of due process, that could not be considered a jury of our peers.

The Compliance Committee may declare a party noncompliant, publicly shaming them without any apparent due process, and tack on an arbitrary additional 30 percent obligation in greenhouse gas reductions. As a truly fascist enforcement mechanism, decisions of the committee usually cannot be appealed.

The Kyoto Protocol jumped from a "we're all in this together" to full on fascism in one leap. It was impressively bold. This was the blitzkrieg of the climate industry power grab.

In 2015 the Conference of Parties (COP), held in Paris, took Kyoto much further. This agreement baked in a five-year cycle when demands placed on the participants would become more ambitious. Like provisions in the Kyoto Protocol, the Nationally Determined Contributions (NDC) seem to come out of thin air, presumably the result of some political bartering outside the public view.

As typical, the United States was burdened disproportionately with finance- and technology-sharing.

President Trump withdrew the United States from the Paris Agreement, citing our disproportionate burden compared to leading polluters like China. He stated in part "The Paris Agreement handicaps the United States economy in order to win praise from the very foreign capitals and global activists that have long sought to gain wealth at our country's expense."

The Climate Change Deception

Of course, it's reasonable to believe that Tom Karl, the director of the National Climatic Data Center, falsified the climate data record to influence the Paris Agreement. He had been a lead author for the IPCC since 1990 and a federal government climate change operative whose work undermined American interests for nearly three decades.

From the establishment of the UNFCCC and the IPCC to the Kyoto Protocol to the Paris Agreement, these international mobsters made an aggressive power play to take over control of the world economy. All of those things were presented to the public as sensible and necessary steps to achieve a sustainable environment. When we look in hindsight, it was clearly a carefully orchestrated power grab.

And what they did to our WDCs was not a coincidence. Abolition of the WDCs to replace them with a politically malleable WDS was a significant part of taking full control of the institutions that control the climate change narrative. Getting rid of the honest and dedicated climate scientists like me was an integral part of that.

After all, we are the ones who know where the bodies are buried.

The Hockey Stick Paper

Much ado has been made about a four-page paper published in *Geophysical Research Letters* (GRL) by Dr. Michael Mann in 1999. The paper looked at paleoclimate data over the past six centuries showing an abrupt increase in global temperatures since the Industrial Revolution.

Mann's hockey stick paper was published just a year after the Kyoto Protocol during a time when the UNFCCC was working toward an eventual implementation of the carbon credit system. It's no surprise that the publication was leveraged to advance the narrative that burning fossil fuels was causing catastrophic global warming. Organizations including the IPCC and the National Academy of Sciences later worked to endorse the core conclusions, while Mann and both his coauthors were awarded with prestigious authorships in IPCC assessment reports.

Their conclusions state in part, "The 1990s are likely the warmest decade, and 1998 the warmest year, in the last millennium."

As a scientist who has published in that journal, I am amazed by the amount of attention it has received. GRL is an important publication hosted by the American Geophysical Union that gives Earth scientists a way to quickly publish a limited set of results. Quick publication helps bring attention to our main findings so the larger scientific community can start broader work on the topic while we work on a more comprehensive presentation of our results. For those of us who are part of the good old boy club or otherwise think our detailed work will pass muster, we might follow up on our GRL article and try to publish our next paper in the more prestigious *Journal of Geophysical Review*.

I wouldn't have imagined that Michael Mann could expect anything more to come of his paper when it was submitted for publication. As far as I can tell, he presented a good manuscript that was well cited, he brought attention to the known uncertainties and important questions, and the manuscript seems to have been peer reviewed by a competent panel of scientists. Even the results that show the "hockey stick" seem to have withstood the test of time and appear to be consistent with the observed rate of global warming. The paper brought Dr. Mann's findings to the attention of the scientific community, helping facilitate future research, exactly as it should have done.

If I were a reviewer, I might have criticized the manuscript for having too much information packed into a four-page paper and told the author to develop it more and submit it to a more appropriate journal. When I read the paper, the presentation of the data analysis was too ambiguous for it to be apparent that the data analysis techniques were legitimate. If I had the foresight to anticipate the controversy it would create, I would have asked the author to more carefully scrutinize the twentieth-century temperature records for inaccuracies. But as far as I can tell, Dr. Mann didn't try to massage his plotting parameters to highlight his findings, which is something that would have been expected and acceptable. I probably would have recommended this paper be published without any major revisions.

That is not to say I agree with or accept the results found in the paper. To the contrary, I believe that any significant anomalies need to be fully scrutinized, and the integrity of the data along with the references he used are fair game for criticism. And that is the point—the article should never have become anything more than a step toward future research.

And that is where Dr. Mann seems to have gone off the reservation of good science. The responsible thing for him to do would have been to acknowledge that his work was a simple GRL paper intended to spur new research in the scientific community. Dr. Mann's paper was important politically, but not very significant scientifically. Instead of exercising an appropriate amount of humility, he went on to publicly claim that the work was mainstream, and he used that claim to parrot the IPCC talking points about how much we must reduce carbon.[39] He went on to leverage the paper's fame to build a career as a political climate change "scientist."

This is how one becomes a regular on the panels that strong-arm governments into terrible decisions. A scientist will generate some public controversy to garner attention and then shamelessly act as a shill for the climate change fascists. They are a small part of the scientific community, but they are put on every panel, so these few who have been bought and paid for can pose as the world's leading scientists. That, at its core, is why climate change "science" pushed by the IPCC and all of their global partners is a fraud.

Some important facts are incontrovertible. With advancements in technology, we have reduced our carbon footprint by remarkable levels.

It is not rational to think that given geopolitics and economic needs, that we're going to reduce it much further. Climate change is probably on a trajectory that is going to come to pass, and only modest corrections might be possible. The idea that we need to spend trillions of dollars on investments that won't significantly reduce our carbon footprint is probably the biggest fraud ever perpetrated on the American people.

And the scientists who have infiltrated our government and our institutions are the ones who are guilty of perpetrating that fraud upon us.

The NOAA Whistleblower

In February 2017 recently retired NCEI scientist Dr. John Bates published an op-ed that blew the whistle on manipulation of the Earth's temperature climate record. This article gained international attention that resulted in congressional hearings,[40] and it brought on personal attacks against Dr. Bates from the climate change community and from journalists.

John's complaint was in reference to an article published by NCEI Director Tom Karl in *Science* magazine. Karl argued there was no warming "hiatus"[x] in recent years. Director Karl's article was published just before an important UNFCCC meeting where the controversial Paris Agreement was adopted. John Bates claimed that the data was manipulated to influence the IPCC and Karl's publication did not meet basic standards required of a peer reviewed scientific paper.

Dr. Bates was an important WDC-A data manager who was known for identifying a most critical issue in archival of scientific data as being "scientists who are unwilling to formally archive and document their data." As a former World Data Center program leader, I agree completely with his findings. In my experience, properly archiving and documenting datasets is a lot of work, and the resources for these activities are very limited. Like Dr. Bates, I also spent a good amount of time cajoling climate scientists to archive and document their data. Throughout my years at NOAA, this was part of my normal work duties. John Bates was a world leading expert on technical issues with climate data.

x The hiatus referred to a temporary pause in the increase of global temperatures.

It's also worth mentioning that there haven't always been well-established methods to archive and document data sets. One of the important accomplishments of the WDCs was to facilitate a community that worked on and adopted sensible data standards. Dr. Bates himself earned NOAA's highest honor for building a revolutionary system to acquire, produce, and preserve climate data records (CDRs).

At NGDC there were established rules and procedures for archiving and documenting data. As a matter of law, data management has been our mandate for decades. According to John Bates, Tom Karl as the director of the combined World Data Centers flagrantly ignored those policies to rush publication of his manuscript. Director Karl was accused of "maximizing warming and minimizing documentation," for the purpose of influencing the IPCC. In other words, Tom Karl, a member of the federal government's Senior Executive Service and a long-time contributor to the IPCC, was directly and publicly accused of leveraging the power and influence of NOAA to influence the obligations that would be forced upon our nation in the Paris Agreement. In the most obvious context, this representative of the US government was being accused of acting against our country's interests.

When investigated, it was found that Director Karls data manipulation algorithms had been destroyed.

Dr. Bates advocated for public access and rigorous scientific standards to be applied to datasets used for scientific publications. But those are the standards that the scientific community had already adopted. As scientists we are expected to adhere to those standards, not only when we publish but also when we review the work of our peers. Sadly, power and influence actively work its way around these standards, and there are many platforms that provide opportunities to undermine them through soft adherence, honor systems, selective peer selection processes, and weakened review processes.

An Implausible Explanation

A key fact is that long-serving deep state government actors hollowed out and repurposed our WDCs. The leadership structure at NOAA suggests that this would have to be done at the level of the assistant administrators, the highest level of leadership that typically survives the presidential appointment process when a new presidential administration takes over. The NOAA WDCs are hosted by the National Environmental Satellite Data and Information Service (NESDIS). The most likely conspirator would be the NESDIS AA, who we know prioritized consolidation of the WDCs, around the same time the international WDC mandate was dissolved.

In 2012 there was a three-month period when incorrect temperature records were published by NCDC and the datasets had to be withdrawn from public availability and reissued.[41] This drew the attention of then NESDIS AA Mary Kicza. AA Kicza was concerned that the errors reflected poorly on NOAA, and she directed improvements to be made with data management. Those changes resulted in increased pressure on data managers like myself to bring our data sets up to par. That was a major pain in the butt because we were already overworked, and NOAA had an ongoing hiring freeze while they were ushering older data managers and program leaders into retirement. What was happening was the AA was working to preserve the reputation of the WDCs while simultaneously working to abolish them.

Those efforts should have made our most publicly scrutinized data like temperature records, more resilient to problems. In 2012 Tom Karl was director of the National Climatic Data Center, and he was responsible for the climate data record that was erroneous and needed to be corrected. So why in 2015 was he circumventing NOAA policy and established peer review procedure by publishing an analysis of a highly circumspect dataset? Even if he wasn't already in the hot seat, the new data handling procedures should have eliminated the possibility of accidental destruction of the algorithms that he used to manipulate that data.

The fact is that in 2015 as NCEI director, Tom Karl published a paper that relied on global temperature records that were manipulated with algorithms that were later destroyed and can never be scrutinized. This happened a few years after NCDC, directed by Tom Karl, released faulty temperature values that embarrassed NOAA and resulted in changes to NOAA's data handling policy. Tom Karl published his paper a few months after being appointed director of the newly merged data centers so he could get the paper published ahead of the Paris Agreement. A preponderance of direct and circumstantial evidence suggests that his actions were corrupt.

John Bates was right. I just wish he had the courage to bring forth his accusations in 2015 instead of waiting until he was retired and the Paris Agreement had been adopted.

My colleagues at NGDC and I didn't know that one of AA Kicza's top priorities was consolidation of the data centers into the new single NCEI. What we did know is that there was an aggressive effort to get rid of so-called unfunded mandates, including termination of most of the WDC-A data services. The problem is that none of those things were unfunded. We now know that NOAA simply redirected those funds to the CI director.

Those data services were the only real justification for the existence of our organization, and the process of getting rid of them had devastated morale. My coworkers and I knew that our mission had been abandoned, and we didn't know what the plan was. Making matters worse, NOAA leadership intentionally kept us all in the dark about their plans.

After ICSU dissolved the UN institution of the WDCs in 2008, some federal supervisors, like my division chief Bill Denig, tipped us off that NOAA leadership was using that to assume our unfunded international mandates no longer existed. It didn't occur to us that the NOAA WDCs were established by an act of Congress, not the United Nations, and we accepted that as a logical, although disturbing, explanation. None of us imagined the NOAA WDCs were being gutted so NGDC and the others could be repurposed as a shell organization for nongovernmental special interests to use to leverage the power and influence of NOAA.

We had no idea that Tom Karl manipulating the CDR to influence the IPCC was part of this larger scale, long-term effort, spearheaded by senior leadership within NOAA. We didn't think much of it when the WDCs merged into NCEI in 2015 and Tom Karl was appointed to be the organization's first director.

An Inconvenient Truth

Much of the whistleblower op-ed by Dr. Bates focused on the process of how data is quality-controlled. This is a dry subject, analogous to how sausage is made, that could bore the most dedicated of us nerds into a coma. The bottom line is that data management is hard, and we scientists do our best to get it right. In seventeen years at the NOAA WDCs, I have never met a low-level scientist who appeared to intentionally falsify a data record. The problems came from "leadership" and those who were seeking leadership positions.

Dr. Bates's article also made some very important points about how the federal government works. He described how the SES bosses do whatever they want and that lower-level employees and managers do not dare question them. The simple fact is if you stand up to leadership at NOAA, they will destroy your career like they have done to me.

Another interesting thing he pointed out is that the younger scientists didn't typically know about archive requirements. This surprised me because the archive requirements are part of the fundamental structure of data center policy, and they are even pressed upon us by the scientific community when we try to publish in any reputable journal. Why would that not be the first thing that new hires are trained for?

But NOAA contends that Tom Karl followed the rules and there was no way to foresee and prevent his computer from blowing up and destroying his data manipulation algorithms. Of course, that was their position; Karl's job was to leverage the influence of NOAA's temperature record to further the global climate change agenda. What else could we expect from a corrupt organization that is allowed to investigate itself?

The reality is that Tom Karl's computer mysteriously failing, without having been backed up, is straight out of the Washington, DC, standard operating procedure for covering your butt.[42] As an institutional matter, it doesn't pass my sniff test because NOAA data center IT policy has always been to back up government computer systems every night. John Bates was right to infer that Tom destroyed the evidence and then lied about it.

John Bates and Tom Karl both hailed from one of our partner data centers located in Asheville. The National Climatic Data Center (NCDC) has been the authoritative source for temperature data since the International Geophysical Year in the 1950s. Dr. Bates was one of the most highly respected scientists at NOAA and was a recipient of the Department of

Commerce Gold Medal for developing a modern metadata system to manage our climate data records.

Dr. Bates publicly disclosed his concerns in 2017, after an extensive effort to inform NOAA management of problems with our climate data record, including what appeared to be an attempt to influence the IPCC. It was smart for him to avoid retaliation that could have risked his hard-earned pension by retiring before making his disclosure. The very fact that he felt he had to retire in order to speak out tells us a lot about the workplace environment at NCEI. What Dr. Bates didn't seem to understand is the bigger picture about how and why the dataset was being misused.

If you want to know how science is manipulated and suppressed, there are not many better examples than the story that John Bates shared with us. His op-ed expressed his concerns and was well written and to the point. What followed was a campaign of personal attacks against him by the climate change fascists.

Based on my personal experience, I am inclined to believe John Bates. What he accused Tom Karl of doing is consistent with the corrupt behavior of NOAA leadership I have witnessed firsthand. Also, NOAA's explanations don't pass muster. In my experience, NOAA leadership believes that they can get away with anything because they know the climate fascists will come to their aid and attack anyone who dares to question them and because our government institutions are allowed to investigate themselves whenever they are exposed for wrongdoing.

CHAPTER 11

WHAT GOVERNMENT IS AND SHOULD BE

It makes sense to figure out what government is and why it has gone so far off of the rails. It seems to me that we are way beyond the point where any objective and rational person could pretend that the government is functional, let alone defend it.

As Thoreau put it, "For government is an expedient by which men would fain succeed in letting one another alone; and, as has been said, when it is most expedient, the governed are most let alone by it." So, I'll present the objective of this book in Thoreau's words: "I ask for, not at once no government, but at once a better government."[43]

Of course, today's America is very different from the one Thoreau knew. A limited government that approaches the ideal of governing not at all is probably not a practical goal, but a better government is at once and always something we can strive for. This is especially true now when our government is at risk of collapse beneath the weight of its own corruption.

After two decades of service, I know that much of the work the government does is important, even when it is not very efficient or expedient. Like most public servants, I worked hard to give the taxpayers a good value for their money. It took me many years to realize that there are many unintended consequences of good work in government, even when we are doing our best. One example is that the best and most efficient work makes government leadership look good, and that provides a veil of cover for those who abuse the system.

At Once a Better Government

The best way to achieve a better government is to remove the incentives for abuse.

At present, our government institutions are structured so the low-level workers do most of the important work and are not involved in the decision-making process. They rarely know what the senior executives are doing and are almost never aware of bad behavior except when it clearly manifests itself. This is analogous to the compartmentalization of sensitive programs where each office is on a need-to-know basis, largely unaware of how their part serves the whole. For typical government service employees, what the Senior Executive Service does is "above their pay grade." Consequently, the senior executives, or deep state employees, have means to act with impunity as long as they don't draw attention to themselves.

One possible solution is to reformulate the government hierarchy to function more like academic research programs. We could allow government service employees to run their programs in the same tradition as scientific principal investigators run their research program. A scientist has to work within a framework that is set forth and managed by administrators, but the administrators do not direct the science itself. A comparable model for government will remove the senior executives from oversight of the day-to-day operations and prevent them from receiving credit for what our civil servants accomplish. The leadership should be administrators and nothing more and be left without the ability to use the good works of our civil servants as cover for corrupt activities.

It could be argued that we need senior leaders to build programs and collaborate with each other. Government is a behemoth, and perhaps we need hands-on dictators for every fiefdom. That is how the government we know functions, but it is neither necessary nor productive. We do not need government leadership to figure out what their agencies should be doing because Congress has already done that for them. Every government agency already has a clearly defined mission.

It is a fundamental truth of human nature that power corrupts, and people who hold power tend to wield it in improper and ill-advised ways. Most of those who abuse our institutions for their personal gain go to great lengths to hide their bad behavior. We only know how bad some things have become because there has been a recent abundance of historically rare instances where the bad actors make mistakes and expose themselves. In many of these instances their arrogance led to mistakes, and the veil they hid behind was pulled away.

In this regard we should thank these cowards and psychopaths for pulling back the veil of cover. The future of our country depends on them being exposed and defeated.

But we also need to defend ourselves against the ones that are smart enough to avoid exposure. We need to identify where our institutions are vulnerable and correct those vulnerabilities. This is how we will defeat those we cannot catch red-handed.

Returning Government to Its Mission

Corrupt mentality is pervasive throughout the government. The consequence is that our government institutions are performing functions they were never intended or given authority to perform. When we consider that our government exists to act by and for the people, through authority enumerated by laws enacted by our democratically elected representatives, the simple conclusion is that we have a rogue government.

We need to challenge the status quo. We should be afforded the presumption that we are not idiots and demand that those with power do what is required to keep temptation from bad actors. I am not even advocating for bad actors to be punished; I want our institutions fixed so they cannot be exploited in the first place. Government can and must serve its people, and it cannot do that while being molested by the people who we entrust the control of our institutions to.

The same can and should be applied to any form of power and influence that resides within our institutions. The dysfunction of those systems will only be exposed by brave whistleblowers who take a stand for our country.

We should judge all of our institutions with the same level of rigor as I have judged NOAA. We need to accept that corruption in any form can be fatal to our republic, so that we stop being dismissive of the thousands of seemingly minor instances of corruption. Like cancer, even the smallest tumor can be malignant and can become fatal.

Anyone who has not been living under a rock should agree that it is in human nature to seek power and influence. The average citizen who lives paycheck to paycheck cannot easily understand what motivates the rich and the powerful, but they really don't have to. The average citizen only needs to understand and accept that those who wield power and influence are different from us, they are likely to be corrupt, and corruption is an imminent threat to our way of life.

There is no doubt that the United States is the most powerful and influential country in the world. Those who seek to exploit the power and influence of a nation naturally covet our government institutions more than any other. We should agree that our institutions are under constant threat and need protection from all enemies, foreign and domestic, including those who we entrust to safeguard them. It is insane to expect those stewards to protect our institutions from themselves.

Repurposing a Government Agency

Today, there is no social issue more pressing than climate change. No matter how anyone feels about climate change, it is without question the single biggest driving force in geopolitics, trade, innovation, manufacturing, and social engineering. There is no issue that has more influence on how we live our lives and what our children's future will look like. It is common knowledge that the National Oceanic and Atmospheric Administration is the world's most influential government institution on the topic.

Other institutions might seem more consequential in our private lives, but the institutions that influence climate change policy are among the most powerful and most influential in the world. It is plainly obvious that the power and influence wielded by NOAA is the holy grail for the climate change fascists who wish to exploit the American government.

We have established that the deep state is shameless in its willingness to betray our trust, and we have established the motives to exploit our institutions. If the American people require absolute proof to fall out of the sky and hit them on the head that our institutions have been hijacked by people who will go to endless lengths to cover their tracks, then there may not be any hope for our future. Critical thought leads us to ask not if, but how our institutions have been hijacked.

Government is a strange manifestation of every imaginable level of bureaucratic hell. Institutions are created for a purpose but can lose direction when that purpose ceases to exist or becomes absorbed by another agency. When the Twenty-First Amendment repealed Prohibition, the army of federal agents who were chasing down the bootleggers had to find something else to do. When the mission goes away, the establishment remains and is repurposed. The repurposing of a government agency creates a unique opportunity for abuse and exploitation.

After a century, what became of the enforcement mechanisms of Prohibition is in of itself an interesting question that has not yet been sufficiently exposed. We know that our government went from cracking down on whisky to distributing hallucinogenic drugs.[44] Isn't it fair to ask if our FBI beat the Mafia, and then appropriated it for their own purposes? Or are we not allowed to ask questions?

Even if government resources aren't always repurposed, the underlying problem is the possibility that they could be. There are not sufficient safeguards in place to prevent the abusive repurposing of sophisticated and dangerous government resources and tactics. The improper repurposing of a government agency can go on in perpetuity, and that highlights the need to promptly decapitate the beast when the congressionally mandated purpose of a government organization no longer exists.

The NOAA data centers are recent examples of government organizations that have been improperly repurposed. The WDCs were part of a Cold War institution that served a great and noble purpose. That purpose was to allow the Cold War enemies, the USSR and the United States, to exchange nonmilitary scientific data, to advance the greater good for all of mankind. Despite the threat of mutually assured destruction, we found a way to collaborate on those things that promoted basic human survival and dignity. Those data centers became world-renowned in the scientific community as beacons of partnership and hope. With that, they became some of the most influential governmental organizations in the world. That influence is in itself a thing of great value.

It might have made sense to dissolve the World Data Centers. We are living in a digital age, and scientific data doesn't need to be carried across the world by FedEx. Environmental data is loaded into massive government databases within minutes and can be instantly accessed by anyone anywhere in the world. The WDC services that were needed in 1957 became obsolete, and there was a need to shift focus to computer storage and high-speed internet. It's regrettable but not surprising that the World Data System that replaced the WDC was made to be a political beast that doesn't sustain or respect the principles of the WDC.

What remained was a few of America's finest institutions: the National Climatic Data Center, the National Oceanographic Data Center, and the National Geophysical Data Center. Those institutions did not dissolve, and they retained tremendous power and influence that was ripe for the taking.

Hijacking a well-known institution is not an easy thing. The global community can't catch wind of it or the institution will lose its reputation, which is its thing of value. The American public might start asking why rogue government agencies are operating without a formal mandate. Congress might start asking what happened to the laws they passed that created the institution. Even the institution's employees can't know what happened because they have to keep producing the good work that the senior executives need to provide cover for their corrupt activities.

So how is a government institution hijacked? I suppose the answer has to be unique to every institution that has gone through the process. Perhaps one day some social scientists will produce a dissertation that outlines the basic anatomy of the process. What I can offer is the story of how this was done to our great World Data Centers.

A Motive to Hijack Our Data Centers

Before we talk about how the institution was hijacked, it helps to understand why. We already established that NOAA wields tremendous power and influence and is the Holy Grail of government institutions to be exploited. We know that NOAA has the power to influence global climate policy, in turn pulling the levers on trillions of dollars of commerce and the means to regulate the behavior of billions of people. From that it's obvious that anyone interested in power and influence would exploit NOAA if given the opportunity. We also know that NOAA, like any government organization, is run by career employees in the Senior Executive Service, who collectively serve as gatekeepers for who can wield the power and influence of their agency.

These SES, or deep state, employees are the gatekeepers. So, who are they?

Through a series of long-term efforts, the US government has become overwhelmingly populated with people who have aligned themselves with the political left. In recent history, government workers have generally been better off under Democratic Party policies, and it is natural for government employees to identify politically with them. Over time, supervisors tend to hire people who share their own personal values, and most people will hire likeminded people. Under the guise of there not being enough scientists to pick from, NOAA agencies use Direct Hire Authority that allows supervisors to pick who they hire over the most qualified candidates.

The process and results are natural, but it has to be recognized for what it is. Our federal government comprises a disproportionately left-leaning workforce compared to the general population. In a time when our nation is so divided politically, this is a problem.

NOAA is an organization that is composed largely of climate scientists and activists, which is a social group in which conservative minds are nearly extinct. For better or for worse, NOAA has one of the most liberal demographics of any organization in our country. This is compounded by the fact that Republican administrations treat NOAA like a political hot potato and make little or no effort to keep the organization on mission. I served under two Republican administrations and two Democratic administrations. The Democratic administrations brought in new agendas and populated the SES ranks with their people. The Republican president appointed Mr. Magoo to run the Department of Commerce, which oversees NOAA, and left Democratic operatives in their positions.

I don't blame Democrats for this. Republican administrations haven't paid attention to NOAA, and both political parties are guilty of being opportunistic. In other organizations, conservative operatives might be running the show, and in all cases, we can be confident that it's not political partisans but more dangerous special interests that are calling the shots.

Capture of a Government Agency

The federal workforce at NOAA is overwhelmingly liberal, and the senior executive positions were mostly populated through Democratic political appointments. This created an environment where political operatives could easily infiltrate the workplace and go unnoticed. But infiltration is the easy part. There are many safeguards in place to make sure federal employees follow a strict chain of command that comes from the president's authority that was given by the American people. Federal employees are not at liberty to fly to Brussels and cut deals with foreign governments and industries.

The brilliant solution that the NOAA deep state came up with was to replace the army of federal scientists at the World Data Centers with contract scientists from academia. Over the course of a decade or longer, the federal scientists were ushered into retirement, and they were replaced with scientists from the several NOAA cooperative institutes, including at the National Geophysical Data Center, which was partnered with CU's Cooperative Institute for Research in the Environmental Sciences (CIRES).

NOAA deep state leadership essentially bribed the university Board of Regents to allow CIRES a modest amount of administrative office space to facilitate hundreds of faculty scientists to work at the federal offices, without an academic mission. Today, hundreds of millions of dollars a year of NOAA base funds goes to CU, and the university takes a big chunk of that off the top. In return, the CI director is a special-interest operative who does not really answer to CU, and the Board of Regents turns a blind eye to the arrangement. The university's scam is not limited to NOAA; CU also takes in nearly $500 million per year from the National Institutes of Health, and more than $1 billion per year from private investment for related research. This has become infamous for bringing you the mishandling of the COVID pandemic; CU takes in roughly $150 million per year from NASA; and a relatively small amount of funding comes from the National Institute of Standards and Technology (NIST), which can still expect an annual picket party of Boulder hippies that claim 9/11 was an inside job.

The cooperative institutes are a massive and corrupt money laundering operation. It is so immense that I had to fear for my safety and the safety of my family when I stood up to the Goliath. Me sharing my story is no less risky than standing up to the Mafia, and it is pretty obvious that our government and its unethical stakeholders have had American citizens murdered for much less.

At NOAA the power and influence that can be bartered is scientific research, results, and policy. That leads to tainted scientific results that the public is brain-bashed into accepting as unimpeachable. Leveraging scientific institutions can be complicated because every research project historically has a different lead scientist who acts as principal investigator (PI).

To ensure scientific integrity, the PI has to be a dedicated scientist who is responsible for the project and ensures the integrity of the results. That is a hallmark of scientific integrity and academic independence. In science, we don't have a supreme overlord who dictates to each scientist what they must work on and what results they should work toward. It is the scientist, as the person most knowledgeable, who directs the research and conducts the investigation. The PI and the peers who review their work give us confidence to trust the science.

At CIRES, what NOAA funds is essentially hundreds of $1 million-per-year research programs. Those programs comprise hundreds of scientists who are all acting independently and are responsible for preserving the integrity of their research. That granularity is difficult to exploit, so the NOAA deep state devised a novel scheme.

NOAA gave a label to its research funding that went to CU and called it "Task-3." Task-3 money is awarded to a single person, the director of the cooperative institute. By doing that, the NOAA deep state made a single person the PI of hundreds of world-class research programs. In doing so, NOAA betrayed every tenet of scientific integrity, and by taking the bribe, CU has betrayed every principal of academic integrity. One person cannot seriously act as PI for even two respectable programs, let alone hundreds of them.

The reason for this scheme is very simple and obvious. The CI director is a plant of some special interest, and his job is to exploit the power and influence of NOAA and the administration of the federal government.

In the case of CIRES, it seems clear that the director is an operative who is tasked to operate outside the bounds of the rules that constrain federal employees. To control what is within the bounds that constrain federal employees, NOAA senior executives unlawfully take orders from the CI director.

CHAPTER 12

CORRECTING VULNERABILITIES AND FIXING GOVERNMENT

Most of the deep state corruption that allows our institutions to be exploited can be corrected. If we can correct these vulnerabilities and root out corruption, we will have a more efficient and effective government. When the government is more efficient and effective, it will be guided by more honest and better ideas. When that happens, our government institutions will do a better job protecting our country and the planet.

A sensible and simple way to fix government is to have it follow our Constitution. Government should do only what is expressly authorized by the American people, as directed through their elected representatives. Our constitutional republic is imperfect, but it is the best system we have.

There is no legitimate reason for rogue government operations, even in the name of national security. To the contrary, the security of our republic is fundamentally undermined by extraconstitutional government activity. If there is a mission that the government must perform, that should be easy to recognize and be codified into law.

But the government does far more than it is authorized to do, and we need to stop that. We need a review of every government program and function, an evaluation of how it serves the congressionally mandated mission and if and to what extent it is necessary. We then need to go through the very difficult process of removing the cancer. This might seem to be an extraordinarily complex proposal, but really it is not; it would be a methodical, finite, step-by-step process.

From the cancer of rogue government operations arises the possibility of exploitation, and through exploitation the public trust is violated. Those who betray the public trust, along with their enablers, employ a range of methods to avoid scrutiny. Among the most common of dishonest methods is convincing us that we need absolute proof of wrongdoing before we should enact commonsense precautions.

It is insane to argue that systematic vulnerabilities shouldn't be corrected until there is absolute proof they have been exploited. Nobody carelessly commits a crime, expecting to be caught. Most crimes are never uncovered, especially when nobody knows that a crime was committed in the first place. The vast majority of abuses of our government institutions go completely unnoticed, but they are still a problem and a threat. We need to do more than fix the problems that we are able to discover. We need a way to fix the problems that we are clueless about. To do that, we need to assume every vulnerability is being exploited and correct it. As soon as we start arguing about whether or not somebody did something wrong, we lose.

As a society it is unwise for us to invite crime through carelessness, and we need to ask if vulnerabilities exist, not if they have been exploited. After all, we secure our personal valuables to prevent them from being stolen before, not after we've been robbed. So why as a nation would we not take proactive measures to protect our institutions from possible exploitation?

I think it would be prudent to focus on the people who try to distract our attention away from potential vulnerabilities. These are the people who claim everything is a conspiracy theory, fight against taking proactive action without absolute proof of wrongdoing, and try to delegitimize proof of wrongdoing when it's been discovered. These people need to be called out on the obvious; they are either misguided, or more likely, they are intentionally working against our common interests. Instead, we let them shame us for stating the obvious, so they can provide cover for their own nefarious motives.

As a matter of basic human behavior, anyone who says, “There is nothing to see here” is probably trying to distract you from something they don’t want you to see. Common sense tells us that their intentions are contrary to the good of those whom they are trying to deceive.

DC “swamp creatures” who disproportionately comprise lawyers and the special interests they represent, continue to demand absolute proof of mal-intent for everything that raises our concerns. What is remarkable is that the citizenry of our country appears to blindly accept that notion. In a court of law, the preponderance of circumstantial evidence is sufficient to constitute proof. As a scientist whose livelihood depends on critical thinking, I think it seems essential that we pay attention to the obvious.

The Climate Change Industry Scam

No matter what you believe about climate change, it is an indisputable fact that our lives are affected disproportionately by the climate change industry and climate change politics. The motive to exploit climate change is plainly obvious and should not be in doubt to anyone. The question that follows is not whether or not climate change has been exploited, but if our country is vulnerable to exploitation. The answer is a definitive yes. The climate change industry and climate change politics are corrupted, and they are being used to exploit the American people.

The most obvious consequence of that is the existence of a nonsensical industry that is heavily subsidized by the American taxpayer. Based entirely on an oversimplified conclusion that climate change is caused by excess carbon in the atmosphere, a multi-trillion-dollar industry that dominates global politics and commerce has been artificially propped up. That industry affects the lives of every person on our planet and serves no apparent legitimate function. The function it claims to serve is simply a deception.

When all factors are accounted for, the solutions proposed by the government advisory boards of the UN have little chance of reducing the amount of carbon into the atmosphere. To the contrary, the solutions being sought will most likely result in an increase of carbon in the atmosphere. Even if those policies could result in some modest reduction, climate change is already baked into the metaphorical cake, and it would be prudent to accept the adverse effects as a predrawn conclusion and mitigate the consequences.

The evidence seems to demonstrate that the climate change industry actually does the opposite of its proclaimed objective. In reality, the industry hinders the implementation of policies and practices that are tried and true in reducing our carbon footprint or could limit the anthropogenic acceleration of global warming. The industry is more than a hoax; it is probably the greatest scam ever perpetrated against the American public.

In part, the scam is maintained through corruption of science. A most basic function of science is to develop, test, and debunk hypotheses. Most scientific hypotheses are incorrect, and rigorous processes are required to make sure that only good science makes its way into our textbooks. It is the obligation of any scientific investigator to rigorously test their own work, and there is a mandate on the scientific community to use the peer review process to probe scientific results for mistakes. Even after research gets past the scrutiny of peer review, the science can't be trusted unless investigations continue with a willingness to reject old paradigms when their faults are exposed.

But there is a "consensus" that the science is "settled." And with that single claim, any scientist should know the science has been corrupted. Anyone who understands the scientific process knows that only a liar with an improper agenda would ever claim science is settled. When someone tells us the science is settled, what they are really saying is that we are not allowed to question it or test its conclusions.

These false prophets are saying that we need to accept their religion on a faith that is blind to truth. We live in a remarkable time where professional scientists are telling us that we must reject the scientific process or else we are guilty of some sacrilege against science. That is absurd. It is a level of absurdity comparable to burning scientists at the stake as heretics.

Dismantling the Administrative State

I believe that we need to cut the head off of the snake that is the administrative state. To do that we must stop facilitating the capability of our government senior executives to act outside of the mandates of the organizations they run. Some have argued that the only way to do this is to fire everyone in the SES. Maybe there is an easier way to accomplish the same goal.

A lateral shuffling of the administrative state could stop rogue behavior. Something as simple as ordering every senior executive to move to a different agency could purge most of the institutional rot. A lateral shuffle of the administrative state could help terminate established channels of corruption and buy us time to identify and do away with many of the systems' built-in vulnerabilities.

I don't believe that senior executives are mostly bad, but many have been put in positions where they have to act against the common good in order to protect their own interests. A comprehensive shakeup of the executive structure of the federal government would remove most of the bad influence and provide civil-servant-minded people an opportunity to return to service for the greater good. With any luck, many will find ways to resist future temptations and become born-again public servants.

We also need to kick these people off of influential policy-making panels. The working groups that advise the government on how to tackle climate change should not be populated with deep state actors who might be compromised. Let's replace them with young scientists who are ambitious and need the opportunity to establish themselves. The language of our government's National Climate Assessment parrots its international counterpart, the IPCC Assessment Report, because the same cabal of compromised scientists who sit on the IPCC panel are placed into every other climate change working group. That creates an echo chamber that prevents new ideas from being put forth, and it isn't productive.

Then there are many who crave power and will do whatever it takes to protect the fiefdoms they have built within their organizations. A lateral shuffling of the administrative state shouldn't be a one-time fix. Instead, senior executives should be moved around every few years so that they are never able to consolidate the power needed to build and sustain these miniature empires. It should also be common practice for every new administration to populate every allowed senior rank[45] with their own people. Doing away with holdovers will help limit the extraconstitutional policies that cause our government agencies to go off the rails.

Regular reassignment of duty stations is a tried-and-true methodology used by the military. In the military, that policy comes at great cost to our soldiers and their families who often have to move every two years. Children of military families often have to move six or seven times during primary school, often in foreign countries, always having to forge new friendships and let old ones go. With few exceptions, our federal workforce can be reassigned without being subjected to the same hardships that our military officers endure.

Limiting the Power of Senior Executives

Another measure I would propose is an executive order that directs the political appointees of each new administration to engage with and directly evaluate their workforce. This is necessary to make sure their respective organizations are acting within their congressional mandate. One thing that seemed to me to be distinctly absent at NOAA was the concept of public service in the workplace. There were only a few employees who regularly asked what was best for the taxpayer, and I've never seen anyone rewarded for doing that. Every federal employee should be encouraged to reflect on what is best for the taxpayer, and to ask themselves if they are serving within the bounds of the laws that allow their organization to exist.

In my seventeen years at NOAA, neither the AA of NESDIS nor the director of NCEI has ever made their way through the halls and asked for a five-minute presentation from each of their group leaders. They never asked me about my work despite the fact that I was the program leader managing most components of the primary NESDIS mandate under CFR 15 Section 950.5(a)(2). The NESDIS mandate is not so overarching and complex that there is any excuse for the assistant administrator to pay zero attention to a key, congressionally mandated program that was under his direction for nearly a decade. There is only one explanation: Our government senior executives are busy doing things they shouldn't be doing at the expense of the mission.

The people who are running the government don't care what their subordinates are doing, and that has to change. Government leaders should regularly block off days and weeks to interact directly with their people, making sure they are on-mission and making sure they have the understanding necessary to manage their organizations. Is Dr. Steven Volz too good to ask me for five minutes of my time? What the heck is he doing as the head of NESDIS? Surely what he is not doing is effectively administering that multi-billion-dollar organization.

Restoring the Meaning of Service

Surely there is no shortage of people who believe they are acting for the common good. I haven't met many federal employees who didn't appear to be working toward that very goal. But someone believing they are working in the interest of the taxpayer and actually doing that are two very different things.

Most low-level workers have no idea how their work fits into the larger mechanism, and they have little basis for assessing how beneficial their work actually is. The only real guide we have is the laws the American people have passed, through their elected representatives. It needs to impress on employees at every level of government service that they work under that authority and no other and that they need to try to stay on that path. It is the senior executives who understand the bigger picture and are charged with keeping their workforce on track. Simply put, the fact that our government agencies are not currently on-mission is proof that the US government's senior executives have failed, and they have failed across almost all government agencies.

When we put our government back on mission, it will be much easier to identify fraud, waste, and abuse and reduce the overall size of government. This is the best and most efficient way to have a smaller, more efficient, and more cost-effective federal government.

Stopping Infiltration of the Federal Government

Another action I would recommend is the immediate termination of any unnecessary and potentially inappropriate government affiliations. The federal government is full of contract workers who perform important support functions. Those contractors include service personnel, information technology professionals, system operators, and even scientists like me. The percentage of the government workforce that consists of contractors has been increasing for years, and that dependence cannot be easily broken. But those workers should only be interfacing with regular government service employees, and they should never be rubbing elbows with members of the senior executive service.

Moving forward, we should be wary of government contractors. The idea that the government shouldn't be competing with private industry might be legitimate outside the walls of the federal workplace, but not within government agencies. Any role that must be performed within the walls of the federal workplace must be considered inherently federal and should be performed by the federal workforce that does not have a conflict of interest. By definition, contract workers have a conflict of interest and have legal obligations and loyalties to entities other than our country. Contractors working within the federal government are an inherent security risk and should be limited as much as is possible.

Simply starting by getting rid of contract supervisors doesn't seem to be a difficult goal. Why do we grant government affiliations to contract supervisors who have no legitimate business within the walls of the federal facilities? Those unnecessary affiliations are improperly leveraged to interface with federal management and, as I have shown, provide a tempting and unnecessary mechanism for corruption. That creates a massive and systematic breach of national security.

We need a complete and immediate purge of these affiliations, along with stringent oversight of conflicts of interest. This should prevent contractors from interfacing with federal leadership and would help limit improper influence. Federal workers should be taking orders within their chain of command, be focused on their mission, and not be influenced by politically powerful contractors.

This idea is no different than the need to keep lobbyists out of the federal workplace. Lobbyists are allowed to influence our elected officials, which we should all agree is morally questionable and needs serious reform. Despite that, it's understood how improper it would be if a lobbyist could pay off a member of Congress to obtain a federal identification so they could issue direct orders to government agencies. That would have those agencies perform for the benefit of their special interest instead of the mandate given to them by the taxpayer. Yet, that is what is happening.

Dismantling Wayward Government Operations

Another thing we should do is put an end to rogue government operations. To do this, we need to create a task force that tracks down the remnants of defunct government agencies and resources and puts them out of commission. This is a necessary extension of scrutinizing the function of government.

We should also investigate how defunct organizations have been repurposed in the past. That could be a painful exercise that might open old wounds, but it is necessary. Understanding the mistakes of the past and bringing them into the open is an important way to prevent those same mistakes from happening again.

This is a serious problem for our country. Publicly facing agencies that have been repurposed, like NCEI, should be pretty easy to identify and clean up. The problem arises when we start to address shadow agencies and functions that have lost their mandates and have been repurposed. Some of those agencies and their functions, such as the National Security Agency and its mission, were created in the dark and never acted within a public-facing congressional mandate. Figuring out how these complex programs have evolved into their present-day manifestation is a monumental task, and fixing their unintended consequences cannot even be contemplated until that is done.

The Fight Ahead

Our universities are obligated to stand up for free and open speech and the integrity of the academic community. When CU allowed my research program to be attacked and for me to be threatened and eventually terminated, it violated its responsibility to uphold the principles of academic independence and tenure.[46] The university compromised its core principles for dirty money, and that is corrupt.

After dragging me through the mud, CU attempted to extort me into signing a nondisclosure agreement that would have barred me from telling this story. I discussed the consequences of that with my wife, I prayed on it, and I declined their offer.

Our American experiment has come to a crossroads, and we are in a very real war against globalist thugs who want to extinguish our way of life and fleece America's carcass. We have a simple but difficult choice: We can stand by and allow our nation to fail, or we can stand our ground and fight.

The corrupt universities are a primary weapon of America's enemies. Not only are they the battering ram that has broken open the doors of our government institutions, but they are also the staging point for every form of attack against our way of life. Through the universities, our children's education has been bastardized, science has been corrupted, and foreign agents have infiltrated our institutions.

It is a multifront war that is being waged on our country, and we cannot win if we only focus on fixing government corruption. To win this war, the people must take back their government, the professors must take back the universities, and the whistleblowers must sing.

I fight because I need to be able to look my children in the eyes and tell them I did my best. I will not stand by while I thrive off the prosperity that has been won with the blood of American patriots on foreign lands. I cannot betray my principles of honor and integrity just because I am afraid of tin-pot dictators. I dare not refuse the mission that God has given me.

I choose to fight.

APPENDIX A

TIMELINE PROVIDED TO W. ABDALATI, JULY 2, 2021

The following is true and accurate to the best of my knowledge and belief. This partial timeline is presented to help you understand the circumstances related to the DURIP and DARPA project problems. Please keep in mind that this document has been put together in a rush, so much of it might have to be revisited for accuracy and completeness. Also, please note these comments are from my handwritten notes, and mistakes may have been made in copying them over (i.e., everything stated herein should be verified). Note, dates are how email threads are dated in my NOAA inbox, and individual emails within a thread may be on different days. Context may be missing on some items.

For brevity, I will use JJM for Dr. Justin Mabie (myself), TWB for Dr. Terry Bullett (Terry), and NAZ for Dr. Nikolay Zabotin (Nikolay).

There are many instances of productive collaboration and goodwill between NAZ, TWB, and JJM. Some of these may be relevant to the current situation and can be referenced as needed.

In 2017 NAS and JJM had a dispute related to the eclipse project. That dispute was settled without any formal complaints being filed, but it might be relevant to the current situation and can be referenced as needed.

4/11/2018—NAZ notified us that the DURIP for Tierra Del Fuego (TDF) was funded. He stated in part that "good science is coming."

4/19/2018—JJM—NAZ. Beginning with a follow-up on Terry's health (TWB had open heart surgery on 4/19/2018), we scheduled an in-person meeting that took place on 4/24/2018. At that meeting we discussed several topics, including the conditions of my working on the DURIP project. This discussion included my research objectives.

Summary of requirements from JJM agreed to by NAZ: TWB and JJM would install the TDF VIPIR according to TWB's requirements (as outlined in the Statement of Work [SOW]). At the end of the DURIP project, NAZ would be allowed to remove the VIPIR radar electronics, but he would have to leave the rest of the site intact. TWB and JJM would have sufficient time to work to obtain funding for new VIPIR electronics and field site maintenance. I made clear that my next decade or more of primary research objectives were dependent on this agreement working, and NAZ indicated he understood and agreed to that. Much of this agreement was made face-to-face, and there is no record of it, but both JJM and TWB can attest to the fact that the agreement was made (and so further proof should not be required).

5/22/2018—TWB—changes some budget items in DURIP proposal (Jessica Rowell).

8/16/2018—TWB—work with CU procurement (Andrew Kulinski).

8/16/2018—TWB, NAZ, JJM—Email chain discussing using a trailer for TDF construction. Some points:

- TWB wants to accommodate NAZ as much as possible. TWB states we need to look at data to see if NAZ's longer-term objectives can be accommodated.
- I provide an update on my portable space weather observatory, which I understood is the portable platform that NAZ was going to deploy his Transportable Dynasonde System (TDS) (the system is proprietary and called RADIO for Rapid Deploy Ionospheric Observatory).

- NAZ highlighted the importance of the TDF instrument being free of obligations. This demonstrates that we all understood the instruments need to be free of obligations.
- I stated a part of my understanding of our agreement.
- Initial differences on goals start to appear.

8/19/2018—TDF trailer discussion is continued.

- TWB tells JJM he just realized NAZ's emphasis on having to have a portable system. He suggests NAZ has unspoken agendas and has deceived us. TWB tells me he plans to hold NAZ responsible to the requirements of the project.
- I state some of my objections and expectations to keep the site funded indefinitely but retrieve electronics after two years.
- I state some of my research objectives with TDF VIPIR.
- NAZ pushes for a meeting, likely to keep some of discussion off of email record.
- I discuss need to use semitruck trailer for portable system (this is integral to the RADIO).

8/23/2018—TWB expresses his frustration with NAZ to JJM.

- TWB claims to JJM that NAZ has some agenda. He does not know this is related to a DARPA project.
- TWB notes that NAZ is getting an improper kickback through a backdoor payment from the VIPIR radar vendor. NOTE: NAZ as the PI is not allowed compensation under the DURIP. However, he is being paid for Dynasonde installation through a back door with the radar vendor.
- TWB notes that he suspects NAZ is improperly moving in on Brett Isham and Cesar Valladares research interests (they are both colleagues from other institutions).

9/10/2018—TWB, JJM, Robert Livingston, and Dick Grubb—inference is made that NAZ is acting corruptly. NAZ wrote to TWB stating that he had explained his intentions to leverage the DURIP for the TDS systems after TWB had objected. NAZ explains to JJM his solution was not just limited to the enclosure but also included the other components (this was a lie—the agreement was that the radar electronics only would be retrieved, and everything else including the enclosure would remain. TWB acquiesced to having the enclosure retrieved but only after much complaining by NAZ). Both TWB and JJM pushed back on the enclosure being moved and suggested that NAZ buy his trailer as a separate item that would not be deployed to TDF.

8/31/2018—Terry, Jose Hormaechea—Details are discussed on TDF construction.

- NAZ pushed back on the construction plans. Note, he should not have any influence on those plans (TWB is engineer for project). TWB pushes back on requirements because TWB is the expert engineer and NAZ is not an expert at field site development. NOTE: NAZ having no expertise in this matter makes him incapable of making decisions on materials, work orders, safety, or compliance—it was understood the construction portion would be controlled exclusively by TWB and JJM.

9/22/2018—NAZ pushing idea to change tower specification in engineering of TDF system.

9/27/2018—TWB to JJM. TWB notes that CU Engineering states they want to increase their DOD footprint. TWB suggests that may be a potential threat to us down the road. JJM notes to TWB that could lead to us being bullied into how we use our instruments. This is relevant now because there is suspicion that they may be trying to improperly secure control of our instruments to secure future DOD funds.

10/8/2018—TWB to JJM and Jose Hormaechea—on NAZ dishonesty.

10/8/2018—NAZ outright lied to Amanda Palacias by telling her the TDF DURIP was for the purpose of acquisition of a TDS. The DURIP was NOT for that purpose—it was to build a world-class field site (requirements of TWB and JJM) and for NAZ to recover the VIPIR electronics (requirement of NAZ). This was a response to TWB conflict of interest complaint against NAZ (107913307). The TDS was not a part of the DURIP in any way—that was part of an entirely different effort by NAZ that JJM and TWB had no part of.

10/10/2018—TWB to NAZ. TWB identified a "hostile working situation." Pushed back on trailer idea for TDF as being a bad idea.

10/8/2018—TWB to NAZ. TWB points out that the transportable Dynasonde is not part of the TDF DURIP. TWB notifies NAZ that he will build the system in the original project description. NAZ requests an in-person meeting.

11/27/2018—NAZ pushes in person meeting. TWB tells NAZ we are busy.

12/7/2018—NAZ again pushes alternatives to TWB antenna design.

- TWB is clearly frustrated with NAZ. TWB implies safety and mission success are at risk.
- TWB insists his engineering judgement must be respected.

12/24/2018—NAZ requests an in-person meeting. Says he has "good news." Meeting is scheduled for Jan 3, 2019 at 2:00 p.m.

- NAZ states that DARPA grant has been awarded. This is the first time JJM has heard about this proposed project. Other agenda items were up for discussion.

1/3/2019—In-person meeting was confrontational. TWB points out that NAZ actions on TDF constitute fraud.

1/4/2019—Following in-person meeting. NAZ sends TWB and JJM "Procurement Fraud Handbook," apparently in response to the meeting on Jan 3, 2019.

1/28/2019—NAZ email chain on action items.

- TWB presents cost spreadsheet for TDF effort.
- TWB presents schedule for TDF installation.
- NAZ accuses TWB and JJM of not doing our job.
- JJM pushes back on claim and insists that quality of work must be maintained.
- JJM details construction and timeline concerns to NAZ.
- JJM reaffirms TWB health concerns.
- NAZ has inappropriate reply. JJM responds with warning that our future efforts are at risk. NAZ responds accusing JJM of making threat.

- TWB clearly states to NAZ that he will not give any guarantees.
- TWB appears unhappy. States he will not meet NAZ alone because he feels personally threatened.
- TWB implies NAZ is impossible to work with, and he reiterates that he does not give any guarantees ever.

2/24/2019—NAZ indicates to TWB that termination from DURIP was imminent.

- TWB tells NAZ that NAZ is altering the scope of the project.

2/28/2019—NAZ fires TWB and JJM from DURIP project.

11/30/2018—continuation of earlier email thread NAZ to Jose Hormaechea on budget items.

- TWB reminds NAZ about material requirement for tower construction.
- NAZ pushes back on TWB material requirements.
- TWB attempts to politely explain the issue. NAZ pushes back against TWB expert opinion.

3/1/2019—TWB notifies Jose Hormaechea and Diego Janches (NASA) that we are leaving the project.

- D. Janches replied with concern and asked who will build the instrument. Note: It is known in our community that TWB is the only person capable of building the instrument that was promised. This damaged our reputation and ultimately led to Jose Hormaechea and his people on the ground at the proposed field site with no compensation despite much work. (From what we understand, NAZ changed the field site and ultimately will use different people. This likely harms the possibility of working with them in the future.)

3/2/2019—NAZ announces Albin Gasiewski and David Kraft to the team.

3/2/2019—JJM to Albin Gasiewski. JJM writes to notify that NAZ is acting unethically and requests a meeting.

- JJM notes an ethics complaint has already been filed by TWB against NAZ.

- JJM notes fear of retaliation from NAZ.

3/2/2019—JJM and Albin Gasiewski meet in person at Albin's home. We discuss the situation. Albin decides to take no action against NAZ and will continue to work on the project despite my concerns.

4/29/2019—NAZ asks TWB if he can use the NOAA Boulder site to perform tests for the DURIP project that he just fired him from. Some back and forth discussion followed.

5/6/2019—NAZ announces that he thinks a DARPA project is about to be awarded, and he offers TWB and JJM to be involved in that effort.

- TWB tells NAZ that we will allow NAZ to use our data from the 2017 eclipse under the condition that he process the data with Dynasonde for us (that data is personally owned property of TWB and JJM). NAZ agreed to that but never delivered the results.

5/6/2019 TWB tells NAZ that he will not perform any experiments with the transportable Dynasonde system. This appears to be a reference to the DARPA project (in effect declining to work on that project).

- TWB warns NAZ that using the transportable system is a very difficult task.
- NAZ offers JJM to participate in DARPA. JJM declines the offer.

7/10/2019—NAZ tells us he needs continuous data at Wallops and San Juan VIPIR systems.

- NAZ makes improper comment about TWB's proprietary Shuffle Mode. He refers to it as the mode "that you love so much."

8/29/2019—NAZ asks JJM to help with DURIP trailer pickup. This would require getting the trailer he purchased from Florida to Colorado.

- JJM provides complimentary cost estimate as expert in transportation costs.

11/13/2019—NAZ asks TWB and JJM to see his system on top of engineering building roof.

- JJM goes to see it and offers complimentary advice on the project.
- JJM reiterates the importance to have a permanent site at TDF.

2/20/2020—NAZ offers a meeting for potential involvement in DARPA project (indicating there is not yet any commitment)

- NAZ tries to rush abstract submission for DARPA.

3/4/2020 Jonathan Snively (Embry Riddle) writes to tell us the BAA is out for DARPA.

3/10/2020—JJM follow up with J. Snively on possible participation.

- Includes discussion on TDF groundbreaking.
- Discussion on shuffle mode not being available for DARPA due to TWB owning the rights to it.
- J. Snively offers for TWB and JJM to participate in DARPA independently of NAZ.
- NAZ states his intention to run TWB's Shuffle Mode.
- TWB tells J. Snively that he will not work with NAZ.
- JJM offers a solution that does not use TWB's Shuffle Mode.
- JJM reminds NAZ that he defaulted on his agreement on the TDF DURIP project.
- JJM tells J. Snively that NAZ is selling him snake oil and NAZ cannot deliver what he has promised.
- TWB tells J. Snively he is concerned with medical side effects from working with NAZ.
- TWB explains to J. Snively how we were tricked into helping NAZ get the DURIP award, then NAZ harmed our interests by changing scope and eventually firing us from project.

3/27/2020—JJM sends NAZ courtesy message that VIPIR will be in Shuffle Mode to observe the Space-X Dragon rocket launch.

- NAZ protests the mode change because of his DARPA project.

2/15/2021—JJM to NAZ—courtesy notice VIPIR is changing to Shuffle Mode.

- NAZ expressed displeasure.
- Note: TWB and JJM were beginning AFRL project and were under orders to protect Controlled Unclassified Information (CUI).

2/19/2021—NAZ writes to tell TWB that he notices the VIPIR was operating in a bistatic mode. We were under CUI restrictions and could not respond to NAZ.

- TWB asks JJM if I told NAZ about the AFRL project. I told him I did not.
- TWB and JJM discuss the possibility that a security incident might be ongoing.
- TWB to NAZ—TWB tells NAZ that the oblique propagation mode is not working. This was true that a Dynasonde compatible mode was not working, but TWB was not allowed to provide more information under the CUI restriction.
- Numerous pieces of evidence that NAZ is attempting to gather information on a controlled AFRL experiment.
- NAZ sent numerous emails to TWB trying to obtain access. TWB responds with simple "No" to access request as per his CUI restrictions. One reply was that it was experimental data and he doesn't want to collaborate.

2/28/2021—Meet with AFRL PI.

- I tell PI that there might be a security incident. I explain that NAZ seems to be trying to obtain information on the experiment and obtain access to the data. PI orders me to restrict all access to the system and make sure all data collected is secured. PI suggests the possibility of sending federal agents to NAZ residence. The look on his face seemed serious. I told the PI I didn't think that was necessary.
- I made no contact with NAZ as per CUI requirements. System access was restored to NAZ after AFRL control of VIPIR was released. I did not provide NAZ with any explanation of why he was locked out as per my CUI restrictions.

3/8/2021—NAZ to JJM. Asks for help accessing systems because TWB is not responding. I did not reply due to my CUI restrictions.

3/15/2021—TWB to JJM. TWB states that NAZ will receive no more complimentary access to VIPIR systems.

3/16/2021—TWB, JJM, Robert Livingston, Dick Grubb—Discussion on NAZ security issues and harassment.

2/14/2021—Collaboration request from Dr. Jee of Korean Polar Research Institute. TWB, NAZ, and JJM all worked for Dr. Jee on installation of Antarctic VIPIR system.

- JJM and TWB reply to all parties on email.
- NAZ does a reply all but removes JJM from email before sending.
- TWB explains the situation with NAZ to Dr. Jee. (NAZ is not included.)
- NAZ responds to Dr. Jee in what we interpret as unprofessional and insulting manner.

4/10/2021—TWB notifies NAZ that he was locked out previously due to security concerns (as required, TWB did not mention AFRL CUI).

- NAZ demands that he be allowed to participate in the scheduling of VIPIR operation. Note: he has never been allowed this privilege.
- TWB tells NAZ that his access has been complimentary and that he owes NAZ nothing. Tells NAZ to leave him alone.
- NAZ demands access to systems, demands his mode be run on the systems, demands he be informed on the system and he wants a discussion on oblique sounding.
- TWB responds with a simple "No."

6/13/2021—TWB—NAZ. Notifies NAZ that VIPIR will be in experimental mode.

- NAZ tells TWB he needs the VIPIR for his DARPA project and other projects.

6/18/2021—NAZ asks TWB if he can deploy his portable system at the Boulder NOAA site.

- TWB replies with a simple "No."

6/19/2021—NAZ asks JJM if he can help drive his TDS to Florida in the fall.

- JJM replies that he cannot help him with his project.

6/29/2021—NAZ tells TWB he needs VIPIR raw data files.

- TWB notifies NAZ that he cannot restore his access and that he cannot explain why.

APPENDIX B

LETTER FROM W. ABDALATI, DATED AUGUST 3, 2021

DocuSign Envelope ID: 6B06391A-D625-4336-BA5A-9A8E0BFD53F8

CIRES

Cooperative Institute for Research in Environmental Sciences

OFFICE OF THE DIRECTOR • WALEED ABDALATI

August 3, 2021

Dr. Justin Mabie
Employee ID #176447
Research Associate / Research Scientist I
CIRES/NCEI

Via Electronic Delivery

RE: Letter of Expectation

Dear Dr. Mabie,

As you are aware, I was recently charged with navigating a complaint from the College of Engineering that revolved around access to scientific instruments that are the property of CU Boulder. My review of that situation is ongoing, and it has required that I work with you, your supervisor Terry Bullett, RIO, and Engineering to resolve it. In so doing, I've interacted with you multiple times. Throughout our interactions, with the exception of our meeting on July 28th, your behavior has been consistently unprofessional and overall disrespectful toward me, the situation, and the entities involved in resolving this complaint. You have on multiple occasions failed to recognize my authority as Director of CIRES, not responded directly to my repeated inquiries, and have made unsubstantiated allegations against other university employees. This is not the first time I have experienced inappropriate and unprofessional behavior from you. I sent you an email on September 26, 2018 raising concerns about your manner of interaction and articulating my expectations, and your correspondences since then have not been aligned with the expectations I spelled out. As a result of your unprofessional behavior, I am, effective immediately, issuing you a letter of expectation. This letter of expectation is based upon your repeated instances of unprofessionalism as outlined in more detail below.

Specifically, I want to outline the exchanges that demonstrate repeated unprofessionalism and direct insubordination on your part.

- Unwillingness to recognize my authority: In my capacity as CIRES Director, I have a responsibility to CIRES employees, the university, and our granting institutions. That obligation extends to conducting a review of the complaint issued by Engineering. I have tried – and continue to try – to conduct my review of the matter in good faith; however, I have repeatedly been met with resistance and disrespect from you.
    - On June 29, 2021, I sent an email to your supervisor, Terry Bullett, (with CC to you) outlining details of the complaint and asking for his clarification on multiple items. You then replied to that email, in lieu of your supervisor, informing me that all future communication should be directed through you moving forward. This attempt by you to dictate the terms of our communications and my interactions with your supervisor, the Principal Investigator involved in the complaint, was very unprofessional and unwarranted.

DocuSign Envelope ID: 6B06391A-D625-4336-BA5A-9A8E0BFD53F8

- On July 2, 2021, my email to you indicated that I would need to set up a meeting with you, your supervisor Terry Bullet, and others to discuss the complaint. I offered to be as respectful as possible of your field work timelines. Your response on July 6, 2021 instead tried to dictate that the meeting be scheduled at your convenience, along your desired timeline. As your Director and the appointing authority for CIRES, your lack of recognition for my authority and related obligations was both inappropriate and disrespectful.
- In my emails to you and your supervisor on June 29, July 6, and a follow-up on July 7, I repeatedly asked you for responses to questions surrounding use of and access to the scientific instruments in question. As of today, those questions have not been directly answered in any constructive manner. Your unwillingness to respond to very straightforward and simple questions, which forces me to repeatedly ask them, is counter to the behavior I would expect any employee to exhibit in response to a request from anyone in the supervisory, in particular the institute director.

- Lack of professionalism: In your email dated July 2, 2021, you asserted that Dr. Nikolay Zabotin negatively impacted your career, violated agreements, and stole intellectual and real property. Such allegations are concerning and potentially very damaging for the individual involved. Despite my reply of the same date, asking for further substantiation of these claims (and again on July 7), with a goal of better supporting you and your supervisor, I never received any evidence of substantiation from you or your supervisor.

Effective immediately, I am issuing the following expectations for your professional interactions with me and others representing CIRES, CU, or our NOAA partners:

- I expect you to respond to all inquiries directed to you to the best of your ability. If you are unable to answer the questions, please explain why that is the case, or ask for clarification.
- I expect that you will respond to emails from me and other university representatives within 24 hours of receipt of a communication, unless you are inaccessible for personal or professional reasons. If you cannot answer within the 24 hour timeframe, you should acknowledge the email and identify a reasonable response timeline.
- If I am requiring your presence at a meeting organized by me, you will attend or communicate a compelling reason as to why you cannot attend. I will try to work with the schedules of all attendees where possible, but you are not ultimately in control of the timelines for these meetings.
- If you would like to include external attendees in meetings organized by me, you will ask for permission first and not assume you can invite external attendees without prior notification and approval.
- I expect that you will avoid making further accusations or allegations about any university employees without substantiation.
- It is my expectation that you demonstrate appropriate behavior and professionalism in all interactions with me and all CIRES, CU and federal colleagues at all times.

You are hereby notified that any further acts of unprofessionalism and insubordination as outlined above will result in further action up to and including termination of your employment with the University of Colorado at Boulder. Your unprofessional actions as outlined above do not represent the spirit of CIRES, CU or our partnership with NOAA. When interacting with others, I expect all CIRES employees, including you, to be respectful and professional to all individuals with whom you interact. You demonstrated this type of constructive behavior during our July 28th meeting, during which you maintained your disagreements with me, and challenged my assumptions, but you did so constructively and respectfully. This behavior was in contrast to the tone of your emails, and consistent with the expectations I am outlining in this letter. To reiterate, significant and sustained professional behavior, as you exhibited during our July 28th meeting must continue to be demonstrated moving forward.

I hope this information is clear. Please contact me if you have questions about any aspect of this letter.

Sincerely,

DocuSigned by:

06F94314731F464...

Waleed Abdalati
Director of CIRES

cc: Terry Bullett, Supervisor
Barry Eakins, CIRES @ NOAA NCEI Lead
Kelly Leandro, Principal Employee Relations Consultant
Angela Knight, CIRES HR Director
Personnel File

APPENDIX C

T. BULLETT SANCTION LETTER, DATED SEPTEMBER 20, 2021

DocuSign Envelope ID: B6D05208-E364-4E3B-B720-81EF5B3424DD

CIRES

Cooperative Institute for Research in Environmental Sciences
UNIVERSITY OF COLORADO **BOULDER**

OFFICE OF THE DIRECTOR • WALEED ABDALATI

Delivered via DocuSign

Dr. Terrence Bullett (EID 200435)
Research Associate, CIRES/NCEI
University of Colorado at Boulder

September 20, 2021

Dear Dr. Bullett,

By way of this letter, I am, effective immediately, issuing you a reprimand. This letter of reprimand follows the resolution letter I issued to you on August 26th and is based upon continued instances of unprofessionalism and insubordination as outlined in more detail below.

Basis for Reprimand: This action is a result of the fact that, after my having made every effort to work with you toward effective resolution on the conflicts associated with past implied commitments by you on the use and access to the Wallops Flight Facility (WFF) and Puerto Rico (PR) ionosonde systems, you have not only continued to maintain a position that is uncooperative and obstructive, but you have displayed insubordinate behavior by continuing to disregard my express directions in a manner that poses significant risk to CIRES's collegial relationship with the College of Engineering and Applied Science (CEAS).

As it stands, you are still refusing to share use and access to these systems with Dr. Zabotin in CEAS, contrary to my directions to you in the August 26th letter of expectation (also referred to as the resolution letter). This was demonstrated once again in your email of September 17th (in reply to his email from two days earlier) in which you again denied him access. Your chosen course of action in refusing to follow the August 26th resolution letter is not appropriate, is not acceptable, and unfortunately, continues to damage CIRES's relationship with university peers and constituents. As detailed below, with this letter of reprimand I am asking you again to comply with the directions and expectations in the August 26th resolution letter, without delay. Failure to do so by 3 p.m. (MDT) Tuesday, September 21st may result in disciplinary action up to and including termination of employment.

Background Information: Your demonstration of uncooperative and obstructive behavior began with our earliest conversations on the matter (May 20, 2021) in which you were always quite vague and very limited in your responses and acknowledgement of your role in the matter, thus making clear your unwillingness to even try to support the interests of a CU colleague, Dr. Nikolay Zabotin, with whom you have worked collaboratively since 2010, and to whom you have implied a commitment to support on his DARPA project. This project is subject to terms of federal grant funding, and your refusal to permit Dr. Zabotin's access to the system—following your many years of doing so—will likely have negative consequences for the project.

Throughout our conversations with respect to granting him such access, you have continually used phrases like "Why should I…?" and "It's not my responsibility…" These comments do not suggest a solution-oriented approach on your part, nor does it uphold the campus core values and commitment to civility and respectful collegial interaction and discourse. None-the-less, I worked with other campus leadership from CEAS, Dr. Massimo Ruzzene, Associate Dean for Research, and the Research and Innovation Office's (RIO), Dr. Jon Reuter, Assistant Vice Chancellor for Research and Innovation, to find a reasonable resolution that could best meet all parties' needs.

This resolution was captured in the letter that I and Dr. Ruzzene sent you on August 26th. The letter was focused only on completing the work that you implicitly agreed to do in order to honor a commitment to our sponsors and in no way bound you to any future interaction or collaboration with Dr. Zabotin. Despite this, you showed no willingness to make any effort to implement our direction and made no effort to find a solution, until a time at which it was clear I would need to take some action in line with what was spelled out in the letter. The solution you finally suggested on September 2nd (granting only limited administrative access), despite my having tried to work with you since May 20th on this matter, was one you dismissed earlier in the ongoing resolution process, when it was suggested by Dr. Jon Reuter during the August 12th Zoom meeting that included you, Dr. Zabotin, Dr. Reuter, Dr. Ruzzene, and myself, among others. I find it frustrating and detrimental to CIRES's reputation that so much of our colleagues' time and energy was consumed and to then, weeks later, have you suggest a resolution that you had previously dismissed. It feels wasteful and intentionally difficult.

I have worked hard to present you with an arrangement that protects your interests and requires only a small amount of flexibility on your part. Your continued response has been simply to claim that security could be compromised, yet, despite my multiple inquires for you to elaborate upon these concerns, you have not stated any concrete risks apart from the fact that you don't trust Dr. Zabotin and the assertion that a person with root access can do a lot of damage. I understand that concern, but our August 26th letter provided mechanisms for avoiding those risks, and you simply dismissed them. Relatedly, when you emailed the CIRES HR Director, Angela Knight, on August 30th, to indicate you did not plan on signing the August 26th letter due to serious disagreements with its contents, you again did not convey to her any context or specific details for those disagreements, despite her asking you directly on August 31st. The goal in my and Ms. Knight's continued asking for context related to your disagreement was to better understand your viewpoint on any perceived limitations, as well as assess, address, and remedy any legitimate risks or concerns you may have had, yet you continued to decline to comment—it is difficult to give credibility to your assertions of security risks when you continually avoid specifying those risks, which also hamstrings our ability to address them. Of course, should legitimate security concerns arise, you would be encouraged to bring them forward.

Following your email to Ms. Knight, you sent a copy of my August 26th resolution letter at least one colleague associated with WFF, Richard Grubb, who subsequently responded to me on August 30th. While it is your discretion to share your job duties and professional expectations with whom you choose, I do find your action of distributing our letter on this CU business matter to external parties who are not in a need-to-know position as unprofessional, an example of poor judgement, and a demonstration of your overall insubordinate and unwilling approach to work collegially and respectfully with me and your campus colleagues. Rather than address these

DocuSign Envelope ID: B6D05208-E364-4E3B-B720-81EF5B3424DD

issues with me directly, you went to an outside party who had no business need to know this information. My hope would have been that you would have instead exerted effort in engaging in a genuine and candid conversation with me, in which you could have provided specific details and concerns about my letter, and we could have worked collaboratively towards a workable solution. Beyond that, you are always free to use other campus resources, some of which are cited below.

Security Report to NOAA: Most recently, you raised concerns, different from those you raised with me, directly with our partners and sponsors at NOAA in the National Centers for Environmental Prediction in your email of September 3rd to LCDR Bryan Begun, the NCEI Executive Officer. In your email to him you stated:

> *"There is an ongoing security incident where a non-Government person is using a position of authority to forcibly obtain root level computer access for another non-Government person who has previously been identified as a security risk. The computer systems are on US Government property at multiple agencies, and may be connected to US Government networks. There are red flags that raise concerns of possible espionage. Please give me a call on my cell phone xxx-xxx-xxxx.*

This is a serious allegation, even if indirectly made, and it understandably prompted immediate investigation from our federal partners. It is certainly your discretion, right, and ethical obligation to report any sincerely held belief of risk or potential misconduct or security risk; CU in general and CIRES in particular encourage such reporting; however, I also would have hoped that you would have initially raised any internal CIRES business concerns through internal CU/CIRES resources, with either myself, Dr. Ruzzene, and/or Dr. Reuter, or with my supervisor, Dr. Terri Fiez, Vice Chancellor for Research and Innovation so that these offices could help resolve the matter. Additional appropriate campus resources you could utilize include:

- Internal Audit: https://www.cu.edu/internalaudit%20
- The Conflict of Interest & Commitment Office: https://www.colorado.edu/researchinnovation/coi
- Central Human Resources, Office of Employee Relations: https://www.colorado.edu/hr/about-hr/employee-relations
- Campus also provides an anonymous reporting mechanism through the Office of Institutional Equity and Compliance (https://www.colorado.edu/dontignoreit/what-report).

Additionally, while it is in your discretion to report concerns directly to NOAA, I would expect that you understand that the relationship between CIRES and our partner and customer, NOAA, is critical to the success of CIRES and of great importance to NOAA.

On that point, NOAA reported to me (on September 8th) that they found no credible security threat related to the concern that you reported to their systems or national security interests. Moreover, NCEI leadership stated to Dr. Wiedinmyer and me, in a Zoom meeting on September 7th that it was their impression that your September 3rd claim of security risk was an attempt to use NOAA as a lever to prevent an action by CIRES that you did not agree with. This was concerning to hear from our NOAA partners; it does not suggest good faith reporting on your part, and the fact that you gave that impression to the leadership of the organization that your work is intended to support reflects poorly on your professional judgment and the integrity/credibility of your report. I was further told that if your claims were in fact true, then your access would need to be removed, because they could not allow you to be in a position to compromise or enable others to compromise NCEI's systems. Let me be clear, reporting genuine, sincerely held concerns is always encouraged and supported; and is an ethical obligation for every CIRES employee, including myself. However, my hope is that you will understand that, going forward, reporting concerns in bad faith or that give the perception that the report was made to intentionally damage relationships or leverage a certain unobjective outcome undermines the credibility of the report and does not meet the expected conduct of employees per the university Code of Conduct.

Importantly, once the security review was resolved by NOAA on September 9th with no risk or threat found and no basis for not following the direction given, I relayed this to you via email (on September 9th), and yet you still took no action on the instructions as laid out in the August 26th resolution letter. In other words, despite NOAA finding no basis for your security concern, you continue to refuse to share the use of the systems with Dr. Zabotin as directed in the August 26th letter. It was my expectation and hope that you would have reasonably assumed that your concerns had been addressed, once NOAA had concluded their investigation with no findings of misconduct or security risks, and that you could then be assured that you could ethically adhere to these expectations. I am disheartened to find that you continue to disregard my directions and act in an insubordinate manner. Based on your continued refusal to follow the August 26th letter, I feel that I now have no choice but to issue this letter of reprimand.

Concluding Statements and Expectations Going Forward: Over the last three months, I have worked diligently to arrive at a fair solution, working with other campus leaders and seeking their input and perspectives. I have tried to ensure your interests were protected by consulting with Employee Relations and the Office of Faculty Affairs, among other offices. I have supported your efforts on ionosonde operations, putting CIRES funds at risk by funding your SAIC work until the contract challenges could be resolved, and taking the unusual step of agreeing to the consideration of a waiver of the publication restriction that SAIC seeks to impose. I have listened to your concerns and carefully considered them as I have worked toward solutions. Yet every meeting we have had, and nearly every correspondence (including your delayed or lack of responses to some direct questions) has demonstrated an unwillingness on your part to work toward mutually beneficial solutions for all campus partners involved; rather they focused simply on achieving the outcome you think is best for your research and from your perspective. In trying to reach a fair resolution regarding use of the systems at issue, I must consider our commitments to other sponsors, the interests of the university, and the interests of university employees. That is what has driven every action I have taken and suggestion I have made. My efforts, however, are not met with even the slightest expression of interest in cooperation and resolution; instead they are obstructed resulting in ongoing strain with CEAS.

As a CIRES employee, I expect you to do the best job you can in the areas over which you have responsibility and freely express what you think is the right thing to do, even if I disagree. I also support, encourage, and expect you to report sincerely held concerns to appropriate resources, which may include NOAA or the campus resources I have listed above. At the same time, however, I expect that you will (1) do your best to work matters of internal CIRES business appropriately first through your CIRES supervisory chain, (2) work with me in a collaborative and professional manner to find solutions when our perspectives are not aligned, (3) accept my judgement on matters that reach beyond your professional domain and that must consider broader CIRES and CU interests, and (4) respect my authority by implementing actions that I request of you or direct you to do, particularly those described in the August 26th letter. You have done none of these, and in the case of #4, I regard your lack of doing so and the manner in which you have acted as insubordination, ultimately resulting in this letter of reprimand. Consequently, as was made clear in my and Dr. Ruzzene's resolution letter on August 26th, continued obstinance to direction and non-compliance with expectations will not be tolerated going forward.

As a result of your continued refusal to abide by the direction that I and Dr. Ruzzene provided to you, I am hereby issuing you this reprimand, and I am reiterating my expectation that you abide by the direction of our August 26th letter immediately. Failure to do so may result in further action up to and including termination of your employment with the University of Colorado at Boulder.

I hope this information is clear. Please contact me if you have questions about any aspect of this letter.

Sincerely,

DocuSigned by:

06F94314731F464...

Waleed Abdalati
CIRES Director

CC: Christine Wiedinmyer, CIRES Associate Director for Science
Barry Eakins, CIRES / NCEI Lead and Sr. Research Scientist
Arnaud Chulliat, CIRES/NCEI Supervisor and Sr. Research Scientist
Angela Knight, CIRES HR Director
Colisse Franklin, Deputy Assistant Vice Chancellor, RIO
Suzanne Soled, Director of Faculty Relations
Kelly Leandro, CU Principal Employee Relations Consultant
Personnel File

Acknowledgement of Letter of Reprimand

I, Terrence Bullett, have received and understand the Letter of Reprimand dated September 20, 2021.

____________________ ____________________

Employee's signature Date

APPENDIX D

LETTER TO E. KIHN, SENT ON DECEMBER 30, 2022

Eric,

It's been a great honor to serve this organization for more than 17 years. I have never compromised on my dedication to our mission to faithfully serve the taxpayer. I sincerely regret, but have worked hard to correct, the mistakes I've made over the years.

Over the past year, it has been difficult to prevent challenges from becoming problems. I did my best, with measurable success, and I hope that has been noticed and appreciated. I am still dealing with the problem that my career has been irreparably damaged, so I'm writing to ask you to help correct that.

What I am asking is that you hire me into the position that is currently advertised (NESDIS NCEI-23-11719805-DHA). I believe this can have a comprehensive corrective result.

I understand that what I'm asking has been normal hiring practice within DSRC. In 2005 the position I applied for was targeted specifically for Rob Redmon; in 2011, Dave Fischman convinced Sue McLean to give him my position; just a few months ago, SWPC targeted a position for Tzu-Wei Feng, and when her application did not make it past HR, they closed the position and reopened it a few weeks later to hire her. The current NCEI position is a DHA[xi] which gives you freedom to choose me as your candidate.

My lawsuit against Abdalati is still an issue that is being managed. The original filing was dismissed without prejudice pending the outcome of the whistleblower investigation. Because I'm still dealing with damage to my career, I will have to file a de novo lawsuit as soon as the investigation is complete. I expect that to be in the next several weeks. The de novo suit has to press claims against all responsible parties, not only Abdalati. If you and I can find a solution now, we can still avoid future litigation.

When the IG investigation is concluded, I understand the recommendations will be sent to the Secretary of Commerce, who will decide what corrective action should be taken. It is very clear that Terry and I were victims of whistleblower retaliation, and the Secretary should act to correct the harm that has caused. I propose the best course of action is to work together to correct that harm preemptively. This would be good for all of us and for the organization.

I really believe that I have continued to grow over the years. The adversity we have faced together has always helped that growth, and I have always tried to learn my lessons and avoid making the same mistakes twice. I was even able to use those lessons to help Bill and Karen through an impossible working relationship, advancing peace and productivity within our unit. A review of the difficult times would show that my actions were always well intended and that I have never taken any action that was not in the best interests of the organization. Here is a brief timeline of how we got to this point.

2005—I applied to a federal position targeted for Rob, and you hired me through CIRES. You told me the position was for Rob, but you would make me a fed if I stuck around for a couple of years. I gave up a good civil service position based on that promise.

2007—You sent me to work with Bill to support WDC as multiple fed retirements were not being replaced. You told me this was my ticket to the federal job you promised.

---

xi DHA is Direct Hire Authority. DHA allows federal agencies to bypass the normal hiring process for specialized positions when there are not enough good candidates. DHA is commonly abused to circumvent federal hiring rules.

2008—I was fully performing as program leader for WDC-A, an inherently federal role. You told me this was my ticket to a federal spot. I helped NGDC deprecate a half-dozen or so federal positions. I performed this role knowing it was inherently federal, but assumed that was okay because management was preparing my federal spot.

2011—My federal position for WDC-A for Geomagnetism program leader opened as promised, but management handed it to Dave Fischman. You told me that Sue demanded he be hired over me, but that he would be moved back to MGG soon and the position would be reopened for me.

2012—Chris Fox retired and I pressed my goodwill with him to push for you to be brought up as his replacement. I told Chris that we needed Eric Kihn because he was a promising leader and the future of our organization.

2015—I was on a 4.5-month trip to Antarctica, living on what was essentially a Korean military base with almost no internet, and during an extremely difficult time for my family. Under your direction, Bill emailed me, telling me the data centers consolidated and I was out of a job.

2019—I was faithfully performing my role as WDC program leader when outside NCEI feds were allowed to come in and run roughshod over my program. At your request, I was fully cooperative and only asked that reasonable steps be taken to protect the data and that I be allowed to use my expertise to help with that. I thought I was acting exactly as you asked me to, but according to Juan, you directed him to threaten to fire me if I didn't keep my mouth shut. This exposed Juan, but I wanted to do the right thing, and I didn't press my whistleblower rights beyond reporting the incident to the IG.

2021—High-level CU administrators conspired with Abdalati to improperly seize Terry's instrument program and hand it to Zabotin, a known security risk. Terry couldn't defend himself, so I stepped in to help find a mutually beneficial solution. In response, Abdalati engaged in a series of threats and attacks against Terry and me.

2021—Terry and I were forced to report that Abdalati was acting suspiciously. You forwarded that report to him, and told him you believed we made it up to avoid having to follow his orders. That gave him the notice he needed to cover up any illegal activity he may have been doing. In response, he retaliated against us.

2022—Whistleblower retaliation sank my federal job applications for the CCOG job that William was given, for 3 positions at SWPC, and for 3 positions at NCEI Asheville.

I hope you will also consider my past performance. When I was appointed to be the first nonfederal WDC program leader, I formally took on the responsibilities of a federal employee. You told me yourself that this position was my move into the federal service and was the fulfillment of your promise. I acted as a federal employee for years, performing the duties of a WDC program leader, which are inherently governmental. Among other things, I was received as an agent of the United States federal government, by high-level representatives of foreign governments, and I negotiated and enforced international agreements on behalf of the federal government.

Petitioning my earned status is not trivial and will open up another messy can of worms. For the common good, I've been withholding filing my petition, hoping we can find another way forward. If you hire me now, I won't have to file that petition; I will start from scratch toward earning my 20 years in service, which is quite a big concession, but it is a way I can show my good faith and gratitude to you for doing the right thing and helping me out of this very difficult situation.

I appreciate your help in putting my career back on track. I am excited to have a new opportunity to serve NCEI and advance your mission. Working together, we can look forward to a happy and prosperous 2023.

ENDNOTES

1. Henry B. Hogue, *Executive Branch Reorganization*, Congressional Research Service, last updated August 3, 2017, https://crsreports.congress.gov/product/details?prodcode=R44909.
2. United Nations, "Hottest July Ever Signals 'Era of Boiling Has Arrived' Says UN Chief," United Nations News, July 27, 2023, https://news.un.org/en/story/2023/07/1139162.
3. David Callaway, interview by Douglas McIntyre, Climate Crisis 247 (website), June 18, 2024, transcript and video, 1:00, https://climatecrisis247.com/news/record-breaking-heat-and-severe-storms-signal-escalating-climate-crisis/.
4. Bastien Dieppois et al., "ENSO Diversity Shows Robust Decadal Variations That Must Be Captured for Accurate Future Projections," *Communications Earth & Environment* 2 (October 7, 2021): https://www.nature.com/articles/s43247-021-00285-6.
5. NCEI, "*Global Climate Report*, National Centers for Environmental Information, 2023, https://www.ncei.noaa.gov/access/monitoring/monthly-report/global/202313.
6. R. Vose et al., "Implementing Full Spatial Coverage in NOAA's Global Temperature Analysis." *Geophysical Research Letters* 48, no. 4 (February 28, 2021): https://doi.org/10.1029/2020gl090873.
7. Julie Kelly, "A Climate Scientist Is Smeared for Blowing the Whistle on 'Corrected' Data," *National Review*, February 15, 2017, https://www.nationalreview.com/2017/02/john-bates-whistleblower-climate-scientist-smeared-global-warming-advocates.
8. Jennifer Schuessler, "Hoaxsters Slip Breastauraunts and Dog-Park Sex into Journals," *New York Times*, October 4, 2018, https://www.nytimes.com/2018/10/04/arts/academic-journals-hoax.html.
9. International Panel on Climate Change, "About the IPCC," IPCC, n.d., https://www.ipcc.ch/about.
10. International Panel on Climate Change, *Sixth Assessment Report*, IPCC, March 20, 2023, https://www.ipcc.ch/assessment-report/ar6.
11. Union of Concerned Scientists, "The IPCC: Who Are They and Why Do Their Climate Reports Matter," Union of Concerned Scientists, last updated October 11, 2018, https://www.ucsusa.org/resources/ipcc-who-are-they.

12. NASA, "Studies Confirm Climate Models Are Getting Future Warming Projections Right," NASA, last updated October 22, 2024, https://science.nasa.gov/earth/climate-change/study-confirms-climate-models-are-getting-future-warming-projections-right.
13. Wikipedia, "James Hansen," last modified April 8, 2026, https://en.wikipedia.org/wiki/James_Hansen.
14. Daisy Dobrijevic and Andrew May, "The Kármán Line: Where Does Space Begin?," last updated November 14, 2022 https://www.space.com/karman-line-where-does-space-begin.
15. Alex Wigglesworth, "A Climate Scientist Wanted to Start a Debate in Academia. He Set Off a Bigger Firestorm," *Los Angeles Times*, September 12, 2023, https://www.latimes.com/environment/story/2023-09-12/california-scientist-blasts-climate-change-research.
16. Thomas S. Kuhn, *The Structure of Scientific Revolutions*, 4th ed. University of Chicago Press, 2012.
17. John Krakauer, *Into the Wild* (Anchor Books, 1997).
18. Justin Mabie, "Rocket Induced High Altitude Acoustic Waves" (PhD diss., University of Colorado at Boulder, 2019), ProQuest (AAT 27662796).
19. M. Zhizhin et al., "Space Physics Interactive Data Resource—SPIDR," *Earth Science Informatics* 1 (2008): https://link.springer.com/article/10.1007/s12145-008-0012-5.
20. International Science Council, "World Data System (WDS)," n.d., https://council.science/member/world-data-system-wds.
21. Kenneth Davies, *Ionospheric Radio Propagation* (US Department of Commerce, National Bureau of Standards, 1965).
22. Don Vaughan, "Movies in Peril: How Restorationists Are Saving Celluloid," *Saturday Evening Post*, January 9, 2024, https://www.saturdayeveningpost.com/2024/01/movies-in-peril-how-restorationists-are-saving-celluloid.
23. Wikipedia, "Cooperative Institute for Research in Environmental Sciences," last modified April 7, 2026, https://en.wikipedia.org/wiki/Cooperative_Institute_for_Research_in_Environmental_Sciences.
24. Wikipedia, "Christine Wiedinmyer," last modified June 6, 2025, https://en.wikipedia.org/wiki/Christine_Wiedinmyer.
25. Mabie v. Abdalati, 1:2021cv02890 (2021), https://dockets.justia.com/docket/colorado/codce/1:2021cv02890/210741.
26. "Newly Declassified Document Indicates FBI Misled Congress on Reliability of Steele Dossier," US Senate Committee on the Judiciary, August 9, 2020, https://www.judiciary.senate.gov/press/rep/releases/newly-declassified-document-indicates-fbi-misled-congress-on-reliability-of-steele-dossier/.
27. "The World Data System (WDS)," International Science Council, n.d., https://council.science/member/world-data-system-wds/.
28. National Archives, "Part 950—Environmental Data and Information," Code of Federal Regulations, September 20, 1979, https://www.ecfr.gov/current/title-15/subtitle-B/chapter-IX/subchapter-D/part-950.

29. National Archives, "Part 1201—Statement of Organization and General Information," Code of Federal Regulations, September 10, 1990, https://www.ecfr.gov/current/title-14/chapter-V/part-1201.
30. Jim Morrison, "The Incredible Shrinking Polar Bears," National Wildlife Foundation, February 1, 2004, https://www.nwf.org/Magazines/National-Wildlife/2004/The-Incredible-Shrinking-Polar-Bears.
31. "The Paris Agreement," United Nations Climate Change, n.d., https://unfccc.int/process-and-meetings/the-paris-agreement.
32. John Bates, "Climate Scientists Versus *Climate Data*," *Climate Etc.* (blog), February 4, 2017, https://judithcurry.com/2017/02/04/climate-scientists-versus-climate-data.
33. "United Nations Conference on the Human Environment, 5–16 June 1972, Stockholm," United Nations, n.d., https://www.un.org/en/conferences/environment/stockholm1972.
34. Wikipedia, "International Geophysical Year," last modified March 10, 2026, https://en.wikipedia.org/wiki/International_Geophysical_Year.
35. Edith Brown Weiss, "Vienna Convention for Protection of the Ozone Layer," United Nations Audiovisual Library of International Law, March 22, 1985, https://legal.un.org/avl/ha/vcpol/vcpol.html; and G.A. Res. 44/228, UN Conference on Environment and Development (March 22, 1990), United Nations Audiovisual Library of International Law, March 22, 1985, https://digitallibrary.un.org/record/82555?v=pdf.
36. "History of the IPCC," International Panel on Climate Change, n.d., https://www.ipcc.ch/about/history.
37. "Treaties: Montreal Protocol," n.d., https://ozone.unep.org/treaties/montreal-protocol.
38. G.A. Res. 44/172 A, Plan of Action to Combat Desertification (December 19, 1989), https://digitallibrary.un.org/record/192783?v=pdf.
39. *Testimony of Dr. Michael E. Mann Distinguished Professor of Atmospheric Science Penn State University and Director, Penn State Earth System Science Center Before the Comm. on Oversight and Reform* (2022) https://docs.house.gov/meetings/GO/GO00/20220208/114392/HHRG-117-GO00-Wstate-MannM-20220208.pdf.
40. Committee on Science and Space Technology, "Former NOAA Scientist Confirms Colleagues Manipulated Climate Records," news release, February 5, 2017, https://science.house.gov/2017/2/former-noaa-scientist-confirms-colleagues-manipulated-climate-records.
41. *National Climate Report* (National Centers for Environmental Information, January 2013), 5, https://www.ncei.noaa.gov/access/monitoring/monthly-report/national/2013/1/supplemental/page-5.
42. James B. Comey, "Statement by FBI Director James B. Comey on the Investigation of Secretary Hillary Clinton's Use of a Personal E-Mail System," July 5, 2016, https://www.fbi.gov/news/press-releases/statement-by-fbi-director-james-b-comey-on-the-investigation-of-secretary-hillary-clinton2019s-use-of-a-personal-e-mail-system.

43. Henry David Thoreau, *Walden; and, Civil Disobedience: Complete Texts with Introduction (*Houghton Mifflin, 2000).
44 Tom O'Neill and Dan Piepenbring, *Chaos: Charles Manson, the CIA, and the Secret History of the Sixties* (Little, Brown, 2020).
45. "PLUM Reporting," U.S. Office of Personnel Management, June 30, 2025, https://www.opm.gov/about-us/open-government/plum-reporting/.
46. "1940 Statement of Principles on Academic Freedom and Tenure with 1970 Interpretive Comments," American Association of University Professors, n.d., https://www.aaup.org/report/1940-statement-principles-academic-freedom-and-tenure.

INDEX